Solid Modeling Using Pro/ENGINEER® Wildfire™

Solid Modeling Using Pro/ENGINEER® Wildfire™

Ajayi O. Adewale, Ph.D., P.E., M.B.A

Associate Professor
Department of Mechanical Engineering
University of North Florida
Jacksonville, FL 32217

THOMSON

DELMAR LEARNING™

Australia · Canada · Mexico · Singapore · Spain · United Kingdom · United States

Solid Modeling Using Pro/ENGINEER® Wildfire™
By Ajayi O. Adewale, Ph.D., P.E., M.B.A

Vice President, Technology and Trades SBU:
Alar Elken

Editorial Director:
Sandy Clark

Senior Acquisitions Editor:
James DeVoe

Senior Development Editor:
John Fisher

Marketing Director:
Dave Garza

Channel Manager:
Dennis Williams

Marketing Coordinator:
Stacey Wiktorek

Production Director:
Mary Ellen Black

Production Manager:
Andrew Crouth

Production Editor:
Andrew Crouth

Art/Design Specialist:
Mary Beth Vought

Technology Project Manager:
Kevin Smith

Technology Project Specialist:
Linda Verde

Editorial Assistant:
Tom Best

Library of Congress Control Number: 2005931264

ISBN: 1–4180–0564–9

NOTICE TO THE READER

CONTENTS

CHAPTER 12 Analysis

Preface

The book is written to quickly acquaint the student with the workings of Pro/ENGINEER® Wildfire™. This is realized in the arrangement of the chapters. Chapter 1 familiarizes the student with key elements of the graphic user interface (GUI) and the use of input devices including the use of a three-button mouse. Chapter 2 gives the student an immediate insight into the methodology of Pro/ENGINEER Wildfire by taking the student through key steps in the creation of a part. This precedes Chapter 3 where Sketcher tools are discussed in more detail. This arrangement is deliberate as it gives the student a quick flavor of the working of Pro/ENGINEER Wildfire and helps to elucidate the place of the Sketcher in Pro/ENGINEER. In Chapter 3, the student also learns more about features of the Sketcher by being introduced to the geometric constraints tools.

Having gained some understanding of Pro/ENGINEER Wildfire, the student is presented with the hands-on workings of revolved features, blends, and sweeps in Chapters 4 and 5. Once the student can create some of the fundamental models, Drawing is introduced in Chapter 6 and Assembly in Chapter 7.

Having provided a quick view of the essential steps on Pro/ENGINEER, some feature operations are covered in Chapter 8. In Chapter 9, the student is introduced to the concept of parent–child relationship and the importance of careful choice of references in part design.

Relations and equations are introduced in Chapter 10. Relations and equations involve some mathematical skills and the chapter is deliberately kept towards the more advanced portion of the book. Chapter 11 introduces parametric design and family table. In Chapter 12, concepts of analysis of models are introduced.

The book is recommended for a practicing engineer who desires a quick introduction to Pro/ENGINEER Wildfire. It is also an excellent tutorial for a CAD course where Pro/ENGINEER may be taught in a single semester. The chapters are written so that the reader does not need a pre-existing file in order to work through any chapter. Nevertheless, anyone without sufficient experience in Pro/ENGINEER is encouraged to go methodically through the entire text, without skipping chapters, in order to benefit fully from the arrangement of the chapters.

Ajayi O. Adewale

October 2005

Dedication

To Folusho, Olufemi and Tolulope

Ajayi O. Adewale, Ph.D., P.E
Jacksonville, Florida (2005)

Acknowledgements

Special thanks go to Toby Meiers for his in-depth and thoughtful technical edit. His knowledge and attention to detail has added a great deal to the book.

I wish to thank Dr. Keith O'Brien for giving me my first industry experience in engineering which opened my eyes to the place of CAD skills in the workplace. My thanks to Dr. Bill Cadwell, Dr. Neal Coulter, and Dr. David Kline for offering me an opportunity in an academic environment which has spawned this work. Finally, I wish to thank the mechanical engineering students at the University of North Florida, Jacksonville, who had worked through the prototype of the book.

The author would also like to thank the following reviewers:

Deanna Blickhan, Moberly Area Community College, Moberly, MO

Peter Fricano, Triton College, River Grove, IL

Shawna Lockhart, Montana State University, Bozeman, MT

Kenneth Perry, University of Kentucky, Lexington, KY.

CHAPTER 1

Introduction

WINDOW INTERFACES

Launch Pro/ENGINEER from the startup menu. The Graphics User Interface (GUI) that opens up will be similar to that shown in Figure 1.1.

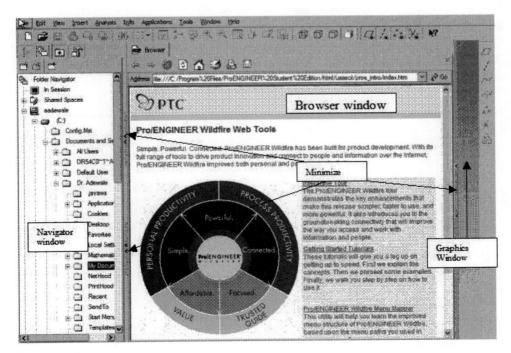

Figure 1.1

At the top of the opened page, we find the usual Windows menu bar and the standard tool bar. Below these are the Navigator window (left), the Browser window (middle), and the graphics area (right). In the Navigator window, we find four options: . The first one is the Model Tree. With no model yet created, it is inactive, but becomes active when a model is created (grayed out). The second option is the Folder Browser. It provides a means of navigating the file system. The third icon provides a means of easily accessing files saved in the favorite section. The last icon is the Connection area that provides access to the browser, team projects, catalogs of existing files, user area and online connection to PTC, as well as to customer support.

Minimize the browser by clicking the minimize button (click anywhere on the line between the short arrows pointing to the left) as shown in Figure 1.1. If you now look to the right, you will see the same two short arrows pointing to the right. Arrows pointing to the right allow you to maximize a minimized window. For practice, go ahead and maximize the browser, then minimize the browser again. You should now be left with a blank graphics window. Created models will appear in this graphics window when opened.

Pro/ENGINEER HELP CENTER

Everybody needs help, so let us get some help with Help. Using the standard menu bar, click **Help → Help Center** . Click Global Search. Type fillet in the **Search for** textbox. The way Pro/ENGINEER Help works is that the database of help files are partitioned according to functional areas. Move **Fundamentals** and **Part Modeling** to the **Selected Functional Areas** using the right arrow button . Your search window now looks like that in Figure 1.2.

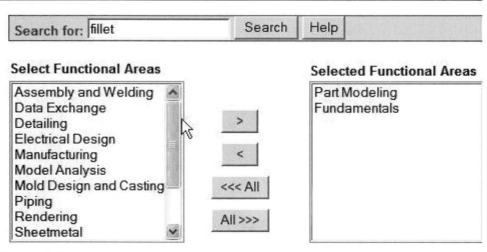

Figure 1.2

Click search and the different help files are displayed. During the actual feature modeling, these help files can be very useful. For now, close the Help windows and return to the graphics window which should still be blank.

ASSOCIATIVE AND PARAMETRIC CONCEPTS

If you have ever used AutoCAD to draw, you realize that once you create a drawing, it is pretty much set and that you cannot easily go back and make changes, unless you re-create the entire drawing. Pro/ENGINEER is a parametric software with bi-directional associativity. This associative program allows you to change a dimension on the 3-dimensional (3D) model and have the 2-dimensional (2D) drawing automatically update with the new dimensions. Since Pro/ENGINEER is parametric, if you want to modify the size of a part, all you have to change is the numeric value of the dimension. For example, if you want to change the height of a boss, then all you would change is the dimension controlling the height. Associated dimensions will correspondingly change. No need to re-sketch anything.

CREATING A FILE

Click **File** → **New** or click on the new file icon ⬜. Pro/ENGINEER opens a "New" file type window, as shown in Figure 1.3.

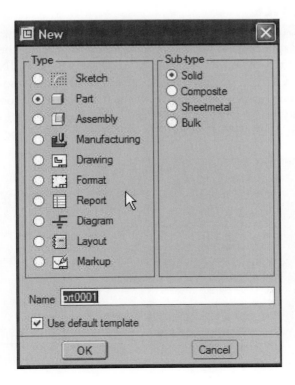

Figure 1.3

Pro/ENGINEER allows several types of file. However, we shall concern ourselves only with the **Sketch**, **Part**, **Assembly**, and **Drawing** options.

Creating a Sketch file allows the user to create a 2D sketch without using any advanced features. A sketch is typically used to represent a cross section, a trajectory, or path. A sketch can be used multiple times within the same part creation. Thus entities like line, rectangle, circle, arc, fillet, spline, coordinate system, point, text, and dimensions are created in the Sketcher environment. The Sketcher environment will be discussed further in Chapter 3.

A Part file is a combination of sketches and other features that create the 3D model. The 3D model typically consists of protrusion, cut, hole, round, chamfer and other features.

An Assembly, as the name suggests, is a collection of parts that are connected with the aid of prescribed constraints.

A Format is a template of the background on which a 2D drawing is placed. Most people create the format once, and reuse multiple times.

A Drawing is typically used to automatically generate a 2D view of a part (3D model). Here is a great benefit of using 3D modeling software such as Pro/ENGINEER – the 2D view representations are automatically generated for you!

OPENING A FILE

Launch Pro/ENGINEER and click on the Navigator tool bar to reveal the folders, just like a Windows Explorer does. Click on the file folder containing the desired file to be opened. Then, right-click on the folder and select **Open**. Pro/ENGINEER displays the files in that folder. Click on any file to select it. Pro/ENGINEER also displays a preview of the model in a small graphics window above the file listings. The tools displayed to the right of the window, as shown in Figure 1.4, can be used to manipulate the previewed file. Click on **Open file in Pro/ENGINEER** icon to open the file in the graphics area.

Figure 1.4

You can also open an existing file by clicking 🖼 and selecting the name of the file from a list.

FILE MANAGEMENT

From the time you first open Pro/ENGINEER, every step is recorded in "Session Memory". This means that if you open a file and then **Window** → **Close** it, Pro/ENGINEER still has it in its memory. Like the active Pro/ENGINEER files, files that have been once opened during a Pro/ENGINEER session remain in the computer random access memory (RAM), even after the files have been closed. Pro/ENGINEER will warn that a file is in session if an attempt is made to name a new file with the same name as the file in "session memory". To get the file from memory, the user will need to carry out the following procedure:

File → Erase → (Current|Not Displayed . . .)

Note that erasing a file does not delete it. Erasing a file simply removes it from the session memory, or from RAM, but still retains a copy of the file in the storage memory like the hard-drive. To physically remove a file from the hard-drive, Pro/ENGINEER provides another mechanism:

File → Delete → (Old Version|New Version)

COORDINATE SYSTEM AND REFERENCES

Pro/ENGINEER does not use the common coordinate system, as used in AutoCAD, for example. Rather, Pro/ENGINEER creates entities relative to established frames of references. For example, Pro/ENGINEER provides a default set of datum planes. Subsequent entities are placed using the distances from these planes. In effect, these planes serve as references. Once an entity has been placed with respect to a reference, this entity is bound to the reference. The entity is dependent (or is a child) of the reference. Conversely, the reference serves as the parent of the dependent entity. This is the basis of inheritance and parent–child relationships (Chapter 9). Entities are related in parent–child referential dependencies. It is this sort of dependencies that make possible the associative and parametric behaviors of commercial CAD packages.

USE OF THE MOUSE

A three-button wheel mouse is a must as several motions of the graphics window can be done by a combination of the keyboard and mouse operations. The following is a summary of the actions that may be accomplished with the use of the mouse. The following abbreviations are useful:

MB – Middle Button of mouse

LB – Left Button of mouse

RB – Right Button of mouse.

These mouse actions are summarized in Table 1.1.

Table 1.1

Spin	Zoom	Pan or move within graphics window
Hold down the MB and move the mouse around. The object on the graphics window spins. The center of spin is the position of the mouse cursor when the MB was clicked, as long as the spin center is off	Rotate the MB wheel or use Ctrl + MB. Hold down Ctrl and simultaneously hold down the MB and drag	Shift + MB

You will get an appreciation of the mouse functionality after we create our first part and then use the mouse to zoom, spin, and pan.

COMMON SHORT CUTS

The **middle mouse-click** ends an operation. It is particularly handy in the sketcher operations, when we want to end an operation and begin another.

Ctrl + A activates a window. This short cut is very useful when several Pro/ENGINEER windows are opened at the same time. May also be useful when Pro/ENGINEER seems not to be accepting commands – window may not be active!

Ctrl + D is used to view a part in the standard orientation.

REMINDER ON *ENTERING* A VALUE

It is very important to understand that to enter a value, as used in this book, means to key in the entry and *actually press the* **Enter** *key*. Pro/ENGINEER will not accept the value until the **Enter** key is pressed.

CHAPTER 2

Extrusion and Holes

GETTING STARTED

In this chapter, we create our first part:

1. Launch Pro/ENGINEER. Minimize the browser by clicking the minimize button at the right-hand side of the browser window.

2. In the Navigator, scroll to and select the installation directory or a desired directory where you want files stored. Right-click on the installation directory and select **New Folder**.

3. Enter *MyFiles* as the name of the new folder. Right-click on the newly created folder and select **Make Working Directory** (this saves all created files in this folder). You can also set your working directory on a different folder, or on your root directory, C:\ for example, by right-clicking on the desired folder and selecting **Make Working Directory**.

You tell Pro/ENGINEER where to save the files, as you create them, by specifying a working directory, typically at the start of a Pro/ENGINEER session. Depending on which version of Pro/ENGINEER Wildfire you are using, the term "Make Working directory" (Version 1.0) is the same as "Set Working Directory" (Version 2.0).

CREATE THE BASE CYLINDER

1. Using the standard menu bar, select **File → New** or click on the new file icon ▢. A window similar to Figure 2.1 opens up.

2. In the **New** window, note that **Type → Part** and **Sub-type → Solid** are pre-selected, as shown in Figure 2.1.

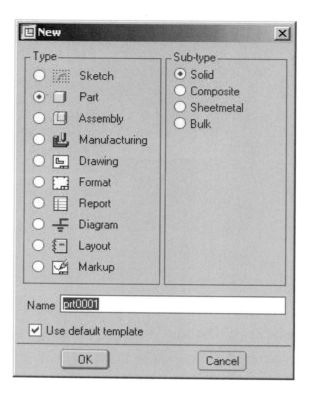

Figure 2.1

3. Enter *cylinder* as **Name** and leave the **Use default template** checked. Click **OK**. Notice that three orthogonal planes labeled **TOP**, **RIGHT**, and **FRONT** appear in the graphics area.

Notice the colored direction indicators at the origin. A plane's direction is uniquely defined by the direction of its perpendicular. The *red* indicator represents the x-direction or the direction of the **RIGHT** plane. The *green* indicator represents the y-direction or the direction of the **TOP** plane. The *cyan* indicator represents the direction of the z-axis or the direction of the **FRONT** plane.

Presently, as shown in Figure 2.2, the graphics window shows that the RIGHT plane is pointing to the right, the TOP plane is pointing to the top and the FRONT plane is normal to the screen. The last correspondence "the FRONT plane is normal to the screen" implies that the normal to the FRONT plane is normal to the screen, which in turn implies that the FRONT plane is parallel to the screen of the computer monitor. (If these

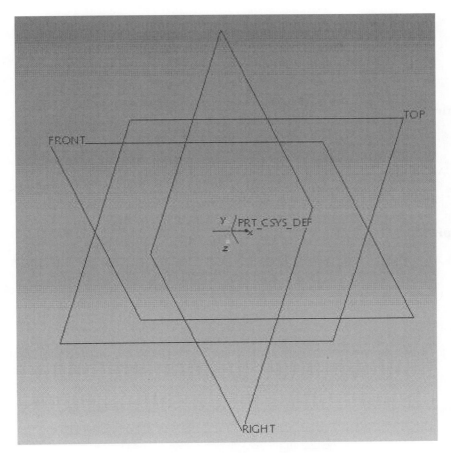

Figure 2.2

planes are not showing, be sure that the Datum planes on/off icon 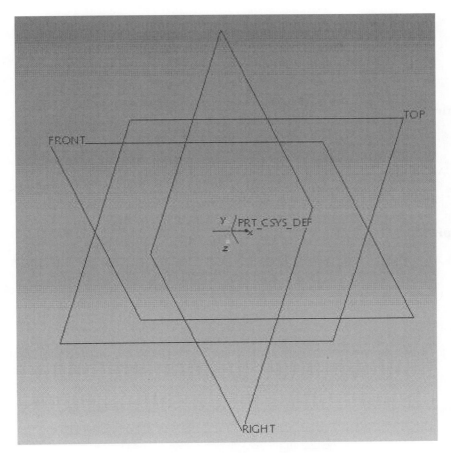 on the standard menu bar is toggled on.)

Take a moment to become familiar with these conventions, as they will become very useful momentarily when we specify the orientation directions. The option to specify the orientation direction offers another flexibility in Pro/ENGINEER as it enables the designer to model a part, viewed from any preferred perspective.

4. Select **Extrude Tool** from the features tools. The features tools are the icons on the right of the screen. Once the Extrude tool is activated, the "dashboard" at the bottom of the screen appears as in Figure 2.3a. The icon for Version 2.0 is shown in Figure 2.3b. The dashboard provides additional options on what the user needs to do.

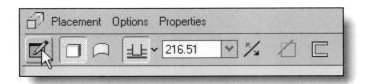

Figure 2.3a

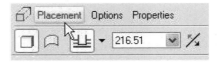

Figure 2.3b

5. Select **Sketch bottom** ☑. This will take us to the **Section** (Version 1.0) or the **Sketch** window (Version 2.0).

   (Version 2.0) Select Placement → Define... .

6. In the **Placement window**, notice the focused box next to **Plane**, as shown by the cursor position in Figure 2.4. This field is ready to be populated by the user.

Figure 2.4

7. Select the TOP plane from the graphics window as the **Sketch Plane**.
    Notice that the datum plane (TOP:F2) appears as the **Sketch Plane**.
    Pro/E assumes which plane would work best for the sketch orientation.

8. Accept the default **RIGHT** plane Reference and **Right** Orientation, by
    clicking **Sketch**, as indicated by the cursor position in Figure 2.5.

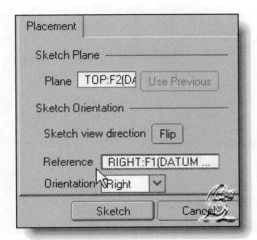

Figure 2.5

Figure 2.5 shows that the feature we are creating will use the RIGHT plane as a
dimensioned reference and the placement on the screen will be such that
the RIGHT plane will be pointing to the *Right*. RIGHT plane as a *reference* means
that the feature's location in space will be tied to the RIGHT plane. Another
way of stating this is that measurements on the feature will be made from the
RIGHT plane. Right plane pointing right on the screen is represented by the red
axes pointing to the right in the graphics window...

9. Click **Close**. You are now in the *Sketcher* environment, as seen in the Figure 2.6a.
    The centered vertical line is the edge of the RIGHT plane while the centered
    horizontal line is the edge of the FRONT plane. The TOP plane is parallel with
    the screen. Pay attention to the illustrations in Figure 2.6b to be sure you clearly
    understand the placement of these planes. Notice the new tools that appear
    down the right-hand side of the graphics area. These are Sketcher tools only
    available when sketching.

10. Select the **Circle Tool** ⭕ . Move the cursor to the intersection of the FRONT
    and RIGHT planes and left-click to specify the center of the circle. Drag the mouse
    to an arbitrary size and left-click. Terminate the circle command by clicking the MB.

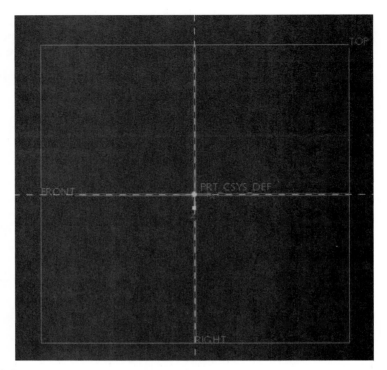

Figure 2.6a

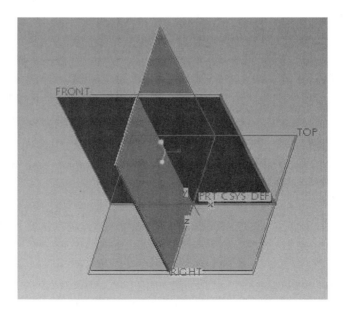

Figure 2.6b

Do not worry about the dimension that was automatically placed for you.

The circle is placed, as shown in Figure 2.7. It is of an arbitrary dimension. Middle-clicking (clicking MB) the second time, after terminating the circle creation, also highlights the **Select Tool** on the right-hand side tool bar. ⬉ As you have noticed, Pro/ENGINEER automatically assigns a diameter dimension to the circle. The dimension of this diameter is considered to be *weak* and is in a gray color. Weak dimensions may be overridden by user-added dimensions.

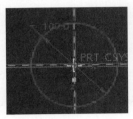

Figure 2.7

11. With the **Select Tool** highlighted ⬉, double-click the weak diameter of the circle just drawn. Enter **10**, as indicated by the cursor position in Figure 2.8. Notice that the circle is resized to reflect the updated dimension.

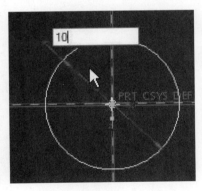

Figure 2.8

Notice that the font of the new dimension is bold. That is because it is user-assigned. User-assigned dimensions are said to be *strong*. Note also that you can change dimensions of a sketch by simply selecting the sketch and dragging it. By dragging, however, specifying exact dimensions can sometimes need more attention compared to just typing in the dimensions, as we have just done.

12. Continue the part creation by clicking the **Complete the Sketch** button ✔. This is the *blue* check mark located on the right-hand side tool bar, below the Sketcher tools. It is a *continue* button, not to be mistaken with **Complete the Feature**

icon, which is green. The green check mark is presently inactive and is discussed more in Step 15. The continue operation makes sense, because the creation of the part was automatically paused when we entered the sketcher to create a sketch of the cross section. We have now sketched the cross section of the part, but we must also specify the depth to create a 3D feature.

13. On the dashboard, enter **20** as the depth of the cylinder (Figure 2.9).

Figure 2.9

14. Click the Preview button represented by the eyeglasses 👓 on the dashboard.

15. Press Ctrl + D to view the object in the standard orientation, as shown in Figure 2.10.

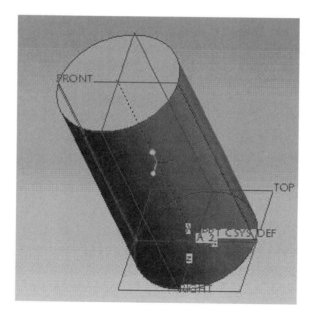

Figure 2.10

16. Turn the middle wheel to refit the object on the screen, as desired. Accept the feature by clicking the green check mark ✔. Notice the completed "Extrude" feature is added to the model tree.

 Remarks: Notice that the top plane of the created cylinder on which we asked the feature to be created, during the initial dialog, is perpendicular to the cylinder at the base and that the direction of extrusion is along the positive green axis (also shown as the positive y-axis). In coordinate geometry, as mentioned previously, the direction of a plane is defined as the direction of its normal. Notice also that the orientation, which we chose to be the right during the dialog, is such that the RIGHT plane is facing the right direction, indicated by the red axis facing the right.

We are now going to create another cylinder, this time the cylinder will be placed on top of the one we just created. An extrusion feature needs a plane on which to sketch its cross section. Our new cylinder is an extrusion feature and needs a plane on which to sketch the cross section.

The top surface of the previous cylinder can be selected as the plane on which to sketch. However, for practice, we are going to create a new datum plane to be coincident with the top of the plane. The method of creating a datum plane can be used in situations where no plane exists. The new feature will be created on a new datum plane on top of the cylinder we have already created.

CREATING THE TOP CYLINDER

1. Use the mouse wheel to refit the cylinder so that it occupies a small area of the screen.

2. Select the **Datum Plane Tool** 🗗 from the right-hand side tool bar.

3. Notice the datum plane creation dialog box appears (Figure 2.11). In Version 2.0, Pro/ENGINEER may assign a reference for you and the space at the cursor position in Figure 2.11 may not be blank. In any case, click on this space so that it has focus.

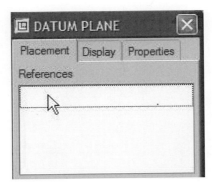

Figure 2.11

4. Click the top surface (*the top surface, not the edge*) of the existing cylinder in the graphics window (Figure 2.12a) and enter **0** as the **Offset Translation**, as indicated by the cursor position in Figure 2.12b.

5. Click **OK**. A new plane DTM1 is added to the Model Tree window, as shown in Figure 2.13.

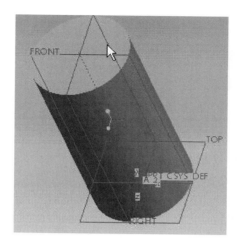

Figure 2.12a

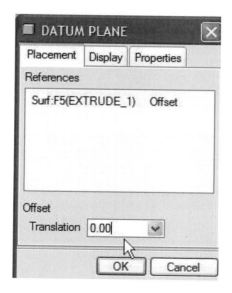

Figure 2.12b

Figure 2.13

6. Click the **Extrude Tool** 🔲, and then the **Sketcher Tool** 🖉.

   (Version 2.0) Select Placement → Define... .

7. In the Placement window that then appears, shown in Figure 2.14a and b, click on the space in front of the Plane to activate the plane selection. (Most times the space, referred to, automatically has the focus in Windows, but the user may have clicked elsewhere to require the step just described.) Then click the

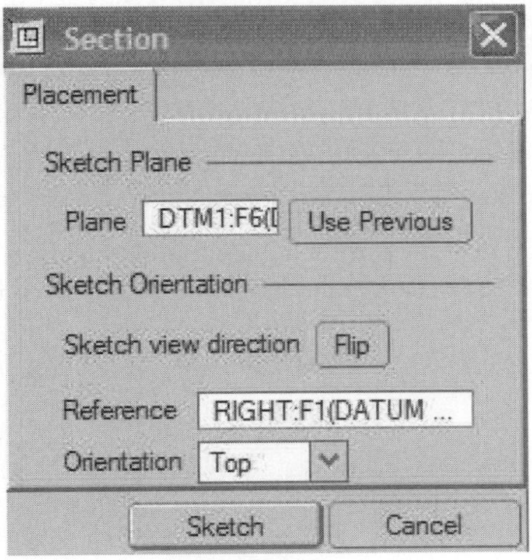

Figure 2.14a

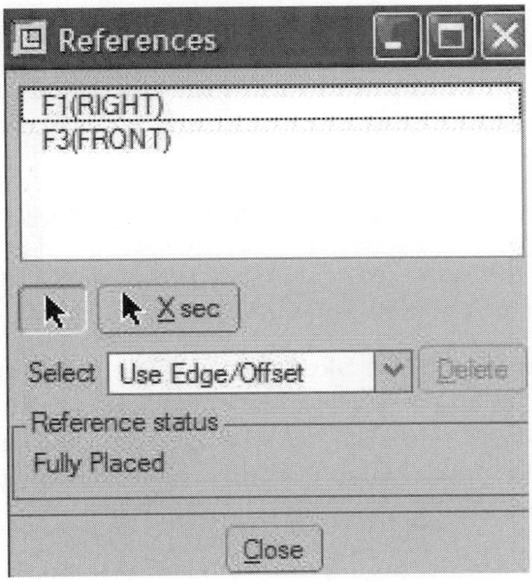

Figure 2.14b

just-created DTM1 plane. (If you have not accidentally clicked elsewhere than described above, the DTM1 plane just created would retain focus and Pro/ENGINEER would already have copied DTM1 as the Sketch Plane for you.) If blank, set the Reference option by clicking the white space in front of Reference and then clicking the RIGHT plane in the graphics window. In the Orientation drop-down box, select the orientation Top, as shown in Figure 2.14a. In Version 2.0, this dialog may be completed for you and you would need to customize it to reflect the setting shown in Figure 2.14b.

8. Click **Sketch**. The **References** window (Figure 2.14b) shows fully placed. The **References** window also shows that the references are the RIGHT and FRONT planes respectively. These mean the feature will be placed relative to the RIGHT and FRONT planes respectively.

9. Click **Close**. The graphics window shows the Sketcher window. Click the **Wireframe** view 🗗 for clearer visibility of the model.

10. Click **Circle Tool** 🔾 and then click the intersection of the FRONT and RIGHT planes to locate the circle center, as indicated by the cursor position in Figure 2.15.

Figure 2.15

11. Drag outwards to create a circle, and left-click when the location of the circle is approximately like that shown in Figure 2.16. Click the MB to cancel the circle command.

12. Click the MB again to activate the **Select items** icon ⬉.

13. Double-click the diameter of the just-created circle and enter **20**, as indicated by the cursor position in Figure 2.17.

14. Click the **Complete the Sketch** icon ✔. In the dashboard at the bottom of the screen, click the depth box and enter **5**, as shown in Figure 2.18. On the right-hand side of the dashboard, click the Preview button 🔬. Press Ctrl + D to view the created feature in standard orientation. Display the model in **Shading mode** 🗗.

15. Click the **Apply and save** button ✔ to finalize the feature creation. The part should now look similar to Figure 2.19.

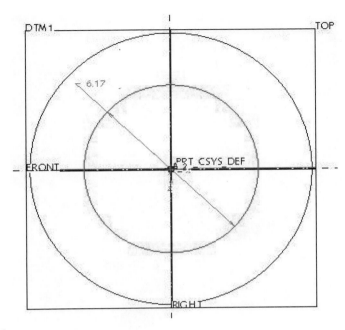

Figure 2.16

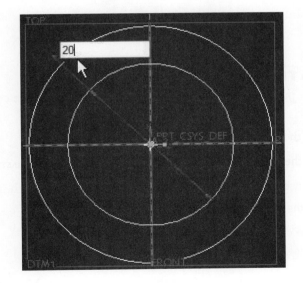

Figure 2.17

Figure 2.18

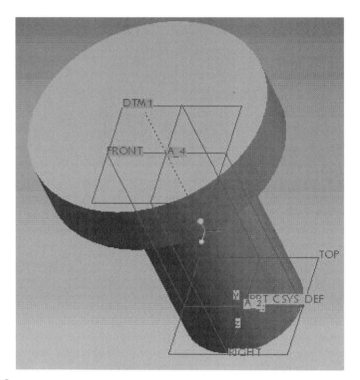

Figure 2.19

HOLE CREATION

We now wish to put a hole on the top cylinder. The hole may be created using one of two methods:

1. Using the extrusion tool with material removal selected.

2. Using the **Hole Tool** or using **Insert** → **Hole** from the standard menu bar.

We are going to use the first method – using the extrusion tool with material removal.

1. Click the **Extrude Tool** .

2. On the dashboard, click **Sketcher Tool** .

   (Version 2.0) Select Placement → Define .

3. In the **Section** window (**Sketch** window, in Version 2.0), click the top of the bigger cylinder as the sketching plane, as indicated by the cursor position in Figure 2.20.

4. Accept the default Reference and Orientation by clicking **Sketch**, as indicated by the cursor position in Figure 2.21.

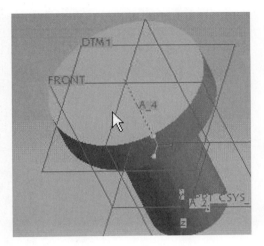

Figure 2.20

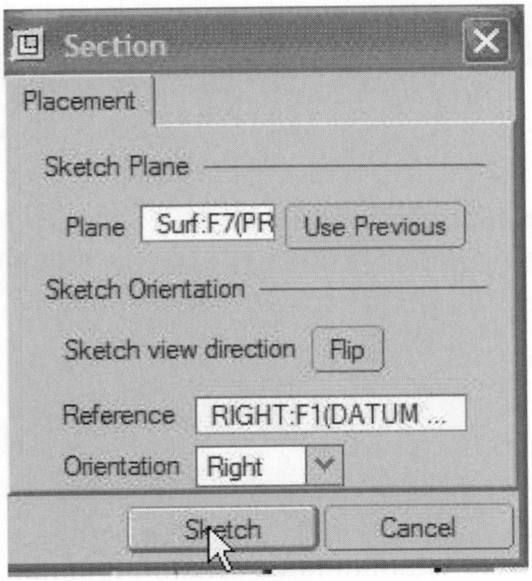

Figure 2.21

5. In the **References** dialog box that appears, click **Close**.

6. Using the technique of sketching a circle, described earlier, sketch the circle of diameter **5** as shown in Figure 2.22 and click **Continue** ✔.

7. On the dashboard, click the **Change Depth Direction Tool** as appropriate to ensure that the direction of extrusion is into the existing model. Ensure that the **Remove Material Tool** icon is toggled on.

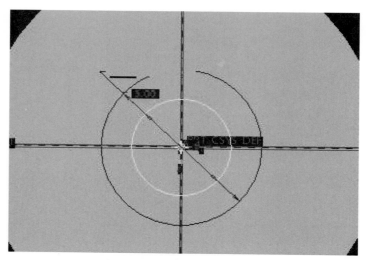

Figure 2.22

8. Preview the feature in the standard orientation: 🔍 → Ctrl + D. If the hole is not created, click Resume ▶ and then click on the **Change Direction Tool** ⚿ again until the extrusion penetrates the existing extrusion.

9. Click the down arrow to the right of the **Depth Tool** ⎯ on the dashboard and select **Through All** as shown by the cursor position in Figure 2.23.

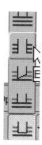

Figure 2.23

10. Preview 🔍 and view in the standard orientation (Ctrl + D). The figure can be spun around to verify that the hole passes through the entire part.

11. Click the **Complete the feature** button ✔. The completed part now looks like that shown in Figure 2.24.

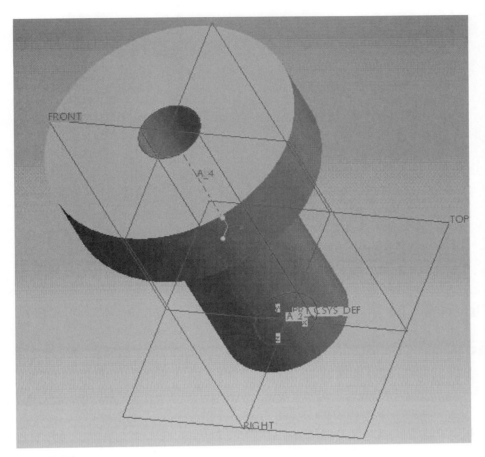

Figure 2.24

CHAMFERING THE PART

We now wish to chamfer the bottom of the lower cylinder and round the top three edges.

1. Click the **Chamfer Tool** (from the set of tools on the right-hand side of screen).

   The dashboard shows the definition dialog for the chamfer.

2. Set D to be 1.

3. With the MB depressed, drag in the Graphics window to reorient the part as shown in Figure 2.25.

4. Select the edge shown in Figure 2.25 — it should turn red.

5. Click the Preview button and then the Resume button.

6. Accept the chamfer feature by clicking the Complete the feature button.

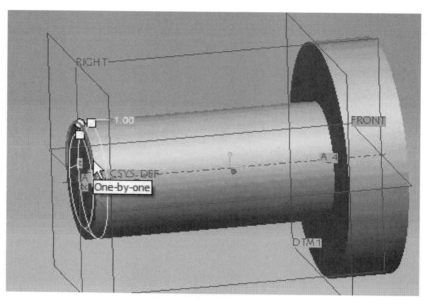

Figure 2.25

ROUNDING THE PART

1. Select the **Round Tool** 🔾.

2. In the dashboard, set the radius of round to **1** (if it is not already set to that), as indicated by the cursor position in Figure 2.26.

3. Select the three edges to round by clicking on them (see Figure 2.27). Make sure to hold the "Ctrl" key between selections.

   • the edge where the two cylinders meet, E1 (+Ctrl)

   • the lower edge of the top cylinder, E2 (+Ctrl)

   • the upper edge of the top cylinder, E3

4. Preview 👓 in the standard orientation Ctrl + D.

5. Complete the feature ✔.

6. **File → Save → OK**.

   The completed part looks like that shown in Figure 2.28.

Figure 2.26

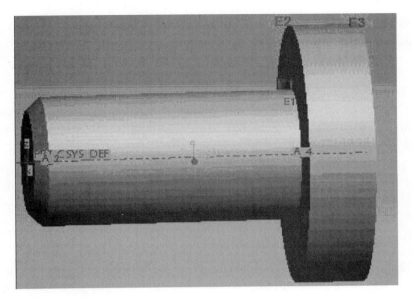

Figure 2.27

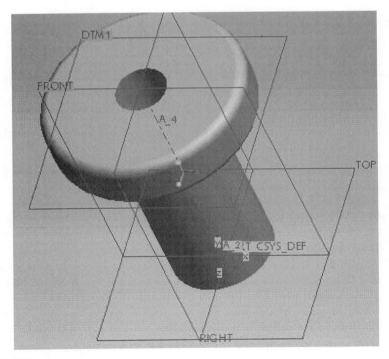

Figure 2.28

INSERTING HOLE USING THE HOLE TOOL

Our task consists of two parts which are described below.

PART (A) – FIRST CREATE A PART

Although we could have used the previous part to demonstrate the hole creation, let us draw another part so as to sharpen our skills about part creation.

In the previous section, we inserted the hole by using extrusion with material removal option. In this section, we are going to create a hole by using the Hole Tool. These are the steps:

1. Start a new part by using **File** → **New** or by clicking the **Create a new object** icon ▢ on the standard tool bar.

2. In the new window, ensure that **Part** and **Solid** options are selected. Leave **Use default template** box checked. Name the part *placed_hole* and click **OK**.

3. Create a protrusion. (In order to insert a hole, a protrusion must already exist!)

4. Select the **Extrude Tool** ▱ in the features tool bar. Click the sketcher ▱.

   (Version 2.0) Select Placement → Define... .

5. In the **Placement** dialog that then pops up, select the TOP plane in the graphics window, as indicated by the cursor position in Figure 2.29. This specifies that we are going to sketch on the TOP plane.

6. Accept the default reference and orientation. Click **Sketch**. Close the **References** window that then pops up.

7. Create the profile shown in Figure 2.30 in the sketcher, using a combination of line and arc commands.

 Important: For the arcs, select the arc tool, click the ends of the line first and drag the mouse to the right to create an arc whose center is to the left of the arc's chord. Remember to use the middle mouse-click to end a command.

Notice that Pro/ENGINEER automatically assigns faint gray dimensions. They are called *weak* dimensions. These dimensions are the Pro/ENGINEER best guesses about what your design intent is. Later these dimensions are given their intended values by the user, then the dimension color becomes bolder. These dimensions are called *strong* dimensions.

We now wish to modify the sketch we have produced. Although, we can modify the dimensions separately, we can modify one side and have

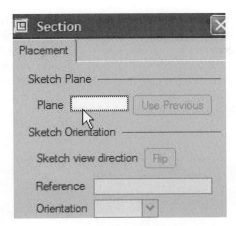

Figure 2.29

Pro/ENGINEER proportionately calculate the rest of the dimensions. This is made possible through the use of the Modify Dimensions tool.

8. Dimension the left vertical line if there is presently no dimension associated with it. Click 📏 and then click the two ends of the vertical line → Middle-click to the left of the vertical line to place the dimension.

9. With the **Selection Tool** 🔺 active, draw a selection window to enclose the created section, by holding down the LB from the upper left to lower right. The dimensions now turn red.

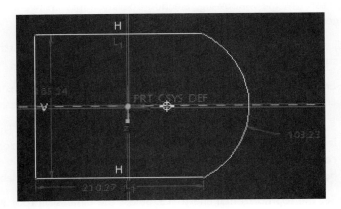

Figure 2.30

10. Click the **Modify Dimensions Tool** 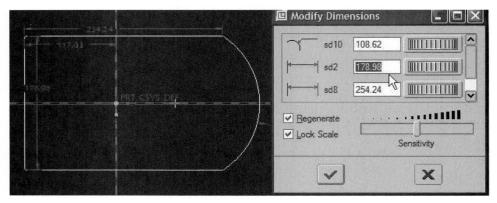. In the **Modify Dimensions** window, check the **Lock Scale** option. In the **Modify Dimensions** window, select the entry corresponding to the vertical dimension. As indicated by the cursor position in Figure 2.31, enter **10** (*Note that enter means press Enter! This scaling process will not work if the user does not press Enter at this stage*). The entire section is rescaled on the graphics window. Click the Regenerate button ✔ in the **Modify Dimensions** dialog box.

Figure 2.31

Notice that the "sd" annotations are a Pro/ENGINEER generated list of entity names. The actual numbers corresponding to the "sd" may vary from screen to screen.

11. If an error occurs, you can undo the work by pressing ⟲ in the standard tool bar. Repeat Step 8 windowing operation, until all the dimensions are selected. Interestingly, the windowing selection operation may change in the selected dimensions for each repetition of the operation.

12. Adjust the rest of the dimensions in the graphics window to match those shown as strong dimensions in the Figure 2.32.

- Ensure that there are no "T" constraints at the intersections of the arc and lines by clicking on them to select → Right-clicking → Delete.

- Select the default dimension of the arc and change it to **6**. The **V** and **H** letters, appearing in Figure 2.32, stand for vertical and horizontal lines respectively. We will cover this in greater detail later. If this operation fails because of other constraints not present in Figure 2.32, delete them.

- Set the top horizontal line to **10**. Select the dimension tool ⊢⊣, then left-click on the left line AB, then left-click on the vertical reference line CD, middle mouse-click to place the dimension. Set this dimension to **4**. Make other dimension changes and delete any constraint not consistent with those specified in the figure. The completed section will be as shown in Figure 2.32.

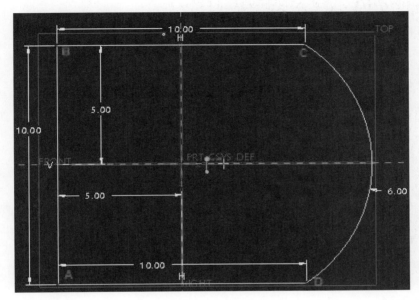

Figure 2.32

13. Click the Complete the Sketch tool ✔. On the dashboard, set the extrusion height to **5** and verify that the dashboard setting is like that shown in Figure 2.33.

Figure 2.33

14. Preview the part 👓 → Complete the feature ✔.

15. Click Ctrl + D to view the created feature in standard orientation. The created feature now looks as shown in Figure 2.34.

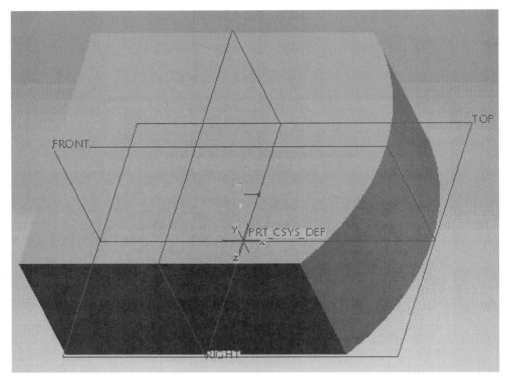

Figure 2.34

PART (B) – NOW INSERT A HOLE INTO THE PART

We are going to place the hole on the top surface of the part.

    I. Click the **Hole Tool** 📌. The left part of the dashboard shows the following options (see Figure 2.35).

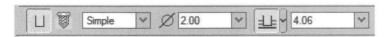

Figure 2.35

The use of these options, like many icons in Pro/ENGINEER Wildfire, may be read by letting the mouse momentarily stay on each of them. Moving from left to right, the first option creates a straight hole, just as the icon shows. The second option allows the hole to be one of the standards. The third option allows use of a simple hole or a sketched hole. The Sketched Hole option enables the user to create a hole which may be wider at the top than at the bottom and can have an irregular profile. The fourth option is the diameter of the hole, the fifth option controls the placement of the hole and the sixth option controls the length of the hole.

Observe that the dashboard, at the bottom of the screen, gives the user the option of selecting the surface to place the hole.

2. Click the top surface of the part, *close to where we intend the hole to be actually located*, for example close to where the cursor is resting in Figure 2.36a.

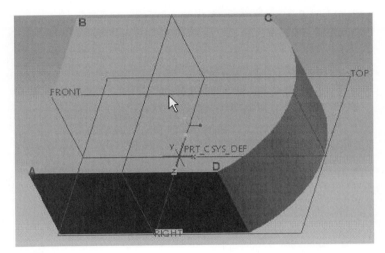

Figure 2.36a

3. Click the **Placement** tab on the dashboard Placement, as shown in Figure 2.36b.

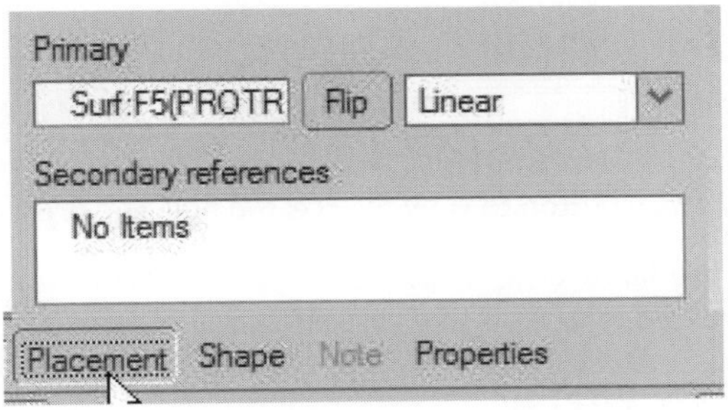

Figure 2.36b

Notice that the Primary is selected (Surf: F5). If the Primary has not been indicated, click the surface ABCD. Pro/ENGINEER needs additional information to uniquely place the hole, so it asks for secondary references. We want the hole to be tied to the locations of the FRONT and RIGHT planes.

4. Click in the selection box under **Secondary references** to place the focus in this box. Holding down the Ctrl key, select the FRONT and RIGHT planes. These selections can be made by either clicking directly on the Graphics window or clicking the listed items in the Model Tree window. Set the offset distances to **0** in each case, as shown in Figure 2.37.
Notice the updated location of the hole with respect to the FRONT and RIGHT planes.

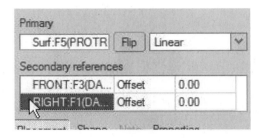

Figure 2.37

5. On the dashboard set the diameter to **4** and select **Through All** for the depth as indicated by the cursor position in Figure 2.38.

Figure 2.38

6. Click the Preview button to view the created hole and press Resume ▶ to continue.

7. Complete the feature ✔. Press Ctrl + D to view in standard orientation, so that the created feature is like that shown in Figure 2.39.

DRILL A SECOND HOLE

Let us now create a second hole, but this time, we wish to place the hole on the side surface parallel to the FRONT plane.

1. Click the **Hole Tool** 🔧 and select the side surface parallel to the FRONT plane as the surface to place the hole, as shown in Figure 2.40.

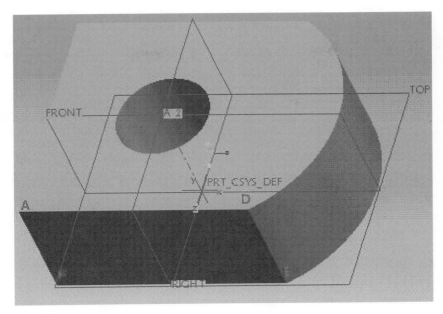

Figure 2.39

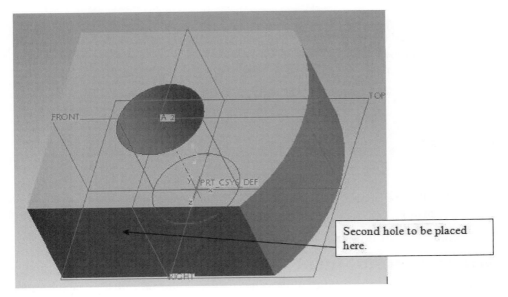

Second hole to be placed here.

Figure 2.40

34

 Note: If you check the graphics window and look at the red coordinate pointer representing the direction of the RIGHT plane, you will notice that it is pointing to the right. Therefore, we are orienting the section so that the RIGHT plane points toward the right.

2. Click the **Placement** tab in the dashboard. Click the white space under Secondary references. Hold down the Ctrl key, select the RIGHT and TOP planes as references. Set the offset for the RIGHT and TOP planes to **0** and **2.5** respectively (see Figure 2.41).

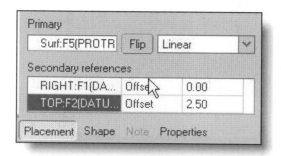

Figure 2.41

3. Set the diameter to **2** and the depth to **Drill up to next surface** (using the drop-down arrow if necessary) as shown in Figure 2.42.

Figure 2.42

4. Click the Preview button and then Resume ▶. Click the check mark to apply changes and complete the feature ✓.

The completed part now looks like that shown in Figure 2.43.

INSERT A THIRD HOLE

Let us now create a third hole. This time, we want the hole to be placed on the curved surface.

1. Click the **Hole Tool** and then select the curved surface (close to where we want the axis of the hole to be placed) as shown in Figure 2.44.

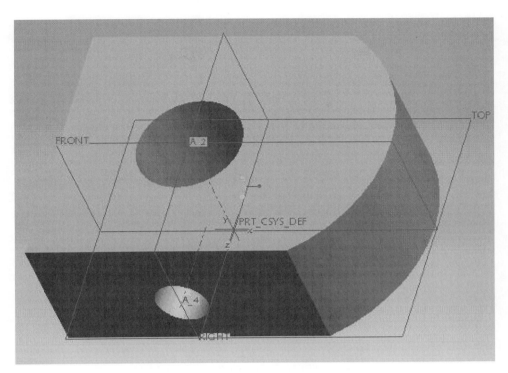

Figure 2.43

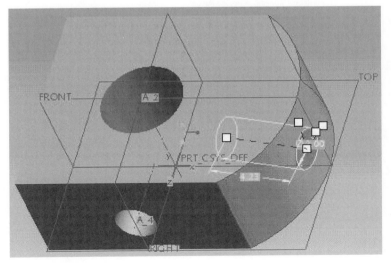

Figure 2.44

We need two perpendicular measurements to serve as references for the hole placement. Pro/ENGINEER needs such information on how to uniquely place the hole, in the absence of the traditional coordinate system.

2. Click the **Placement** tab in the dashboard. As **Secondary references**, holding down the Ctrl button, select the FRONT and TOP planes (after clicking the white space underneath Secondary references, to ensure that this box has focus). Again, selection may be made by clicking in the Graphics window or in the Model Tree window. If an error is made during the reference selection, right-click on the unwanted reference and select Remove. Repeat the steps concerning the reference selection.

3. In the **Placement** window, set the TOP offset distance to **2.5** and the FRONT angle to **0.0** (see Figure 2.45).

4. In the dashboard, set the diameter to **2.0** and the depth to **Drill to intercept with all surfaces** (see Figure 2.46).

5. You may press Ctrl + D to align the view in the standard orientation, as shown in Figure 2.47.

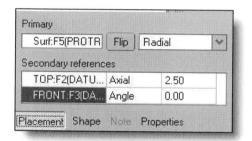

Figure 2.45

Figure 2.46

6. Preview , Resume . Apply and complete the feature .

7. The final part now looks like that shown in Figure 2.48.

We now wish to make a modification to the third hole so that it terminates at the hollow center.

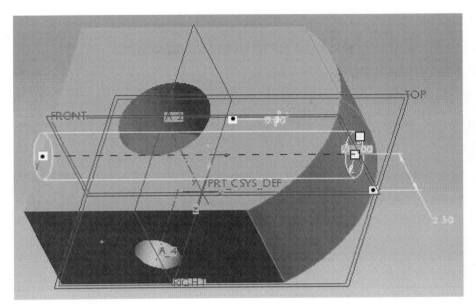

Figure 2.47

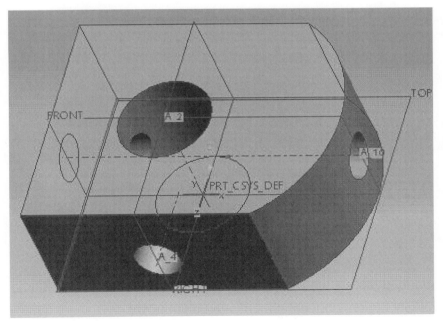

Figure 2.48

1. In the Model Tree, select the third hole and right-click to bring up additional options. Select **Edit Definition**, as indicated by the cursor position in Figure 2.49.

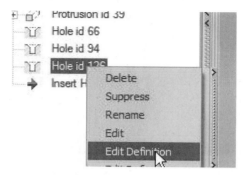

Figure 2.49

2. On the dashboard, click down on the arrow next to the depth option and select **Drill up to next surface** ⬆.

3. Preview, if necessary 👓. Apply changes and complete the feature ✔. The final object then becomes as shown in Figure 2.50.

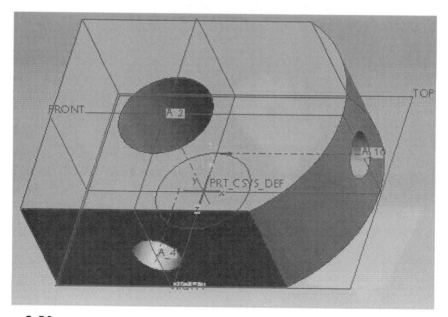

Figure 2.50

SHELLING

We will now *shell* the created part.

1. Click the **Shell Tool** 🔲 on the features tool bar.

2. In the dashboard, enter **0.5** as the Thickness (see Figure 2.51).

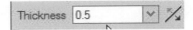

Figure 2.51

3. Click the top surface to be removed to shell this portion of the part, as indicated by the cursor position in Figure 2.52.

   The modification process may not always succeed. It is prudent to always preview before applying changes to a part. This offers the chance to fix any error before applying the changes. Not following previewing first may lead to a difficult resolution step of a feature creation.

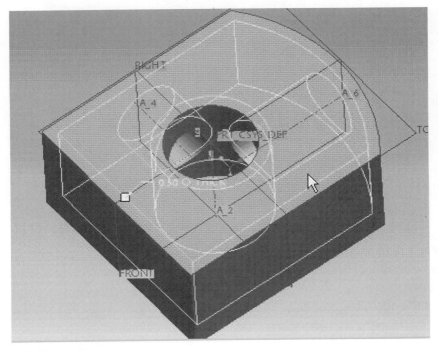

Figure 2.52

4. Click the Preview button 👓. Shelling succeeds.

5. Apply and complete the feature ✔. By depressing and holding down the MB, reorient and spin the model so that it looks like that in Figure 2.53.

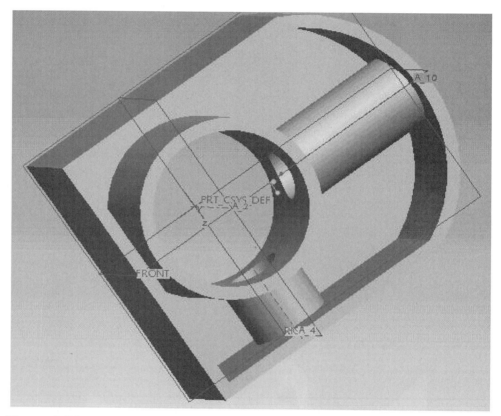

Figure 2.53

SAVED VIEWS

We have on numerous occasions used the standard orientation Ctrl + D. Let us now experiment with the other views.

1. Click the **Saved view list** 🔲 from the standard menu bar. The standard views found in engineering drawing applications are displayed.

2. Experiment with these views by selecting TOP and FRONT from the list (see Figure 2.54).

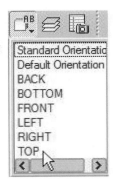

Figure 2.54

First, let us select TOP from the list. The TOP view is displayed. Double-click on the view to display the dimensions and then toggle on the **No Hidden mode** ⬚ for visual clarity. The following dimensioned view (Figure 2.55) is displayed by Pro/ENGINEER.

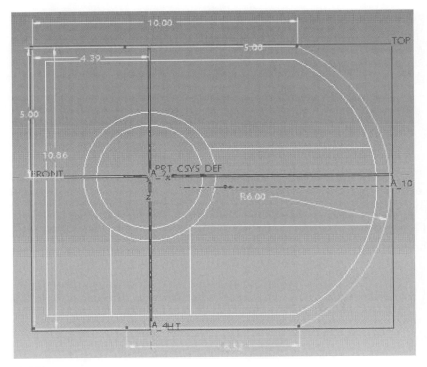

Figure 2.55

Notice the elegance of automatically getting the dimensioned view of a 3D part. More sophisticated 2D drawing examples are covered in Chapter 6.

3. Repeat the step to orient to the FRONT view.

The above are orientations of the viewer. We also have options about the display as demonstrated below.

1. Return to the standard view Ctrl + D.

2. Select the **Wireframe view** 🔲.

3. Select the **Hidden Line view** 🔲.

4. Repeat for the **No Hidden view** 🔲.

THE MODEL TREE

Notice that the model tree on the left-hand side is now populated with the features we created (Figure 2.56).

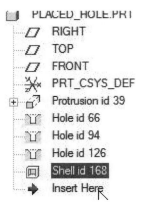

Figure 2.56

Recall that, prior to part creation, the **Model Tree** icon 🔳 was inactive, since there were no features to display at that time.

1. Return to the standard orientation (Ctrl + D).

2. With the *placed_hole* displayed, notice that there are three holes enumerated in the Model Tree.

3. By clicking on each hole icon from the Model Tree, notice that the corresponding hole is highlighted in the graphics display.

4. Click twice on each hole id in the Model Tree and rename them *TOP_HOLE*, *FRONT_HOLE*, and *RIGHT_HOLE*, respectively. Notice that a valid feature name

cannot have space between the name characters and that the feature name is automatically converted to upper case letters.

EDITING A CREATED FEATURE

1. Right-click on the TOP_HOLE feature and then click **Edit**, as indicated by the cursor position in Figure 2.57.

Figure 2.57

The dimensions of these features are shown for editing in the graphics area.

2. Click the diameter value and change it from **4.0** to **6.0**, as indicated by the cursor position in Figure 2.58.

Nothing seems to have changed, except for the color of the dimension – the feature view needs to be regenerated for the view to reflect the changes.

3. Click the **Regenerate Model** icon ⬚ from the standard tool bar.

4. The diameter of the TOP_HOLE is regenerated, as shown in Figure 2.59. Behold the power of an associative CAD package like the Pro/ENGINEER!

The above feature editing enables the user to edit the entities. More fundamental changes may involve creating new entities altogether. For example, the user may wish to convert one of the straight left ends to a curved end. Since the straight end of the part is an original protrusion whose cross section was sketched, a revised sketch is called for and the user needs to go to the Sketcher environment to edit as necessary. We shall now do this next:

1. Right-click on the extruded protrusion icon and select **Edit Definition**, as indicated by the cursor position in Figure 2.60.

Notice that the entire protrusion is selected.

2. The dashboard then opens up. Click the sketcher ⬚.

(Version 2.0) Select Placement → Define....

Click **Sketch** to enter Sketcher. Observe that the sketched profile of the protrusion now opens up, as shown in Figure 2.61.

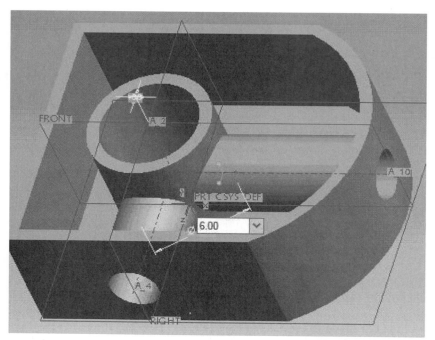

Figure 2.58

3. With the **Selection** icon activated, select the left vertical line so that the line turns red. Then right-click (on the left vertical line) and select **Delete**.

4. Using the **Arc Tool** , click the two open ends of the sketch and drag the mouse to place a semi-circle (of radius **5**) centered as shown in Figure 2.62.

5. Complete the sketch creation by clicking the **Continue Tool** ✔.

6. Click **OK** in the Sketch dialog box that now opens up (upper right corner).

7. Click the Preview button in the dashboard. Click Ctrl + D to view the feature in standard orientation. The recreated feature is now displayed in the graphics window.

8. Click the Resume button ▶.

9. Click the Complete the feature icon on the dashboard ✔. The modified part now appears as shown in Figure 2.63.

    Again watch for the power of associativity at work, as the entire part updates!

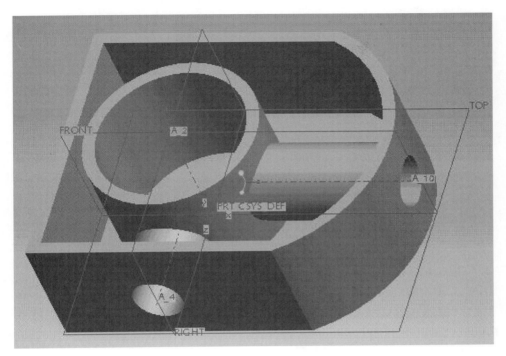

Figure 2.59

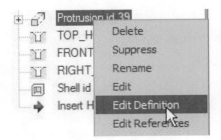

Figure 2.60

10. Click **File** → **Save a Copy**, as indicated by the cursor position in Figure 2.64a. Name the modified file PLACED_HOLE_MODIFIED, as shown in Figure 2.64b.

11. Click **OK**.

12. Close the window.

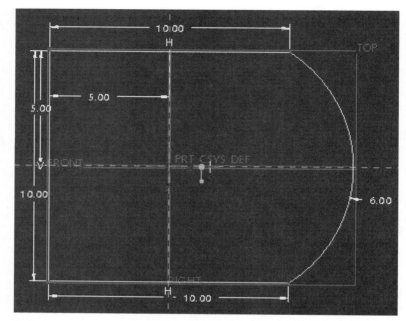

Figure 2.61

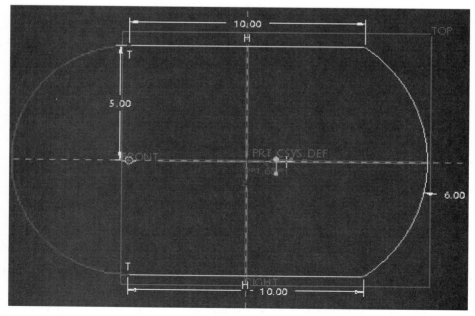

Figure 2.62

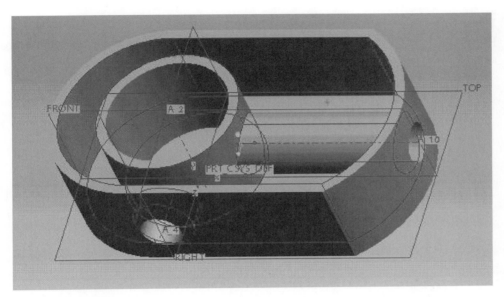

Figure 2.63

Figure 2.64a

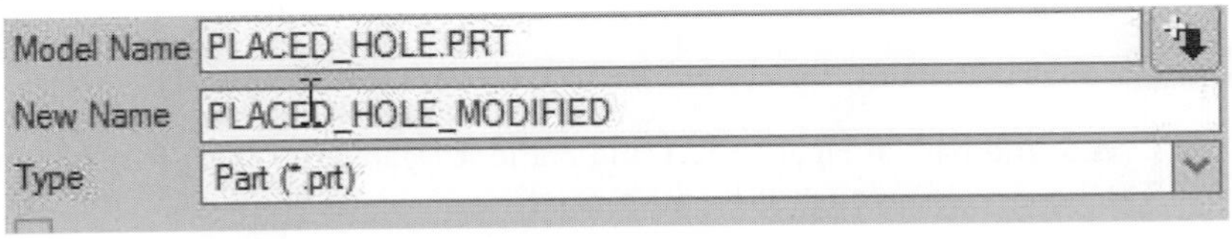

Model Name	PLACED_HOLE.PRT
New Name	PLACED_HOLE_MODIFIED
Type	Part (*.prt)

Figure 2.64b

THE MODEL PLAYER

Pro/ENGINEER provides a very useful tool – the model player – for reviewing the steps in the creation of a part.

1. Toggle **on** the **Shading mode** 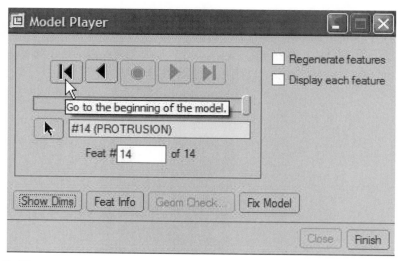, making sure the datum planes, axes, and Coord Sys are also displayed.

2. With a part still opened in the graphics window, using the standard menu bar, select **Tools → Model Player**.

   The following interface opens up (see Figure 2.65).

Figure 2.65

3. Go to the beginning of the part creation by clicking ◄◄.

4. Go through the steps by clicking ►. This step will show you how the entire 3D model was created.

EXERCISE

1. Create the part in Figure Q2.1 and name it *FigureQ2_1*. Use free dimensions. The hole goes through the entire part.

2. Create the part in Figure Q.2 and name it *FigureQ2_2*.

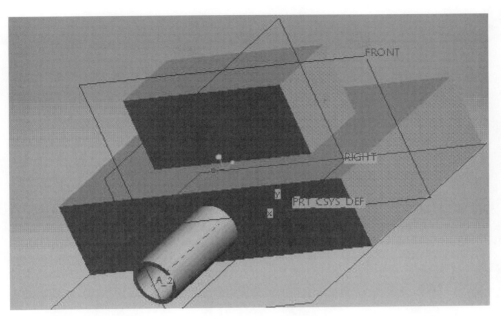

Figure Q2.1

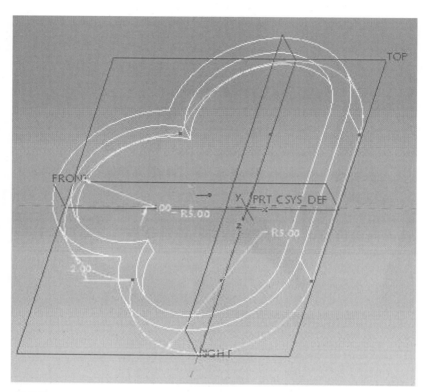

Figure Q2.2a

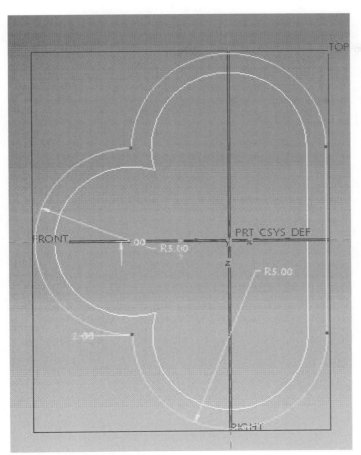

Figure Q2.2b

CHAPTER 3

Design Intention, the Sketcher, and Geometric Constraints

THE BASICS

As you may already have recognized from the previous chapters, Pro/ENGINEER adopts what looks like a "sketch and adjust" philosophy – the designer first sketches a profile that looks similar to what he really wants, then goes on to adjust the dimensions to match the actual intent. The Pro/ENGINEER functionality that makes this possible is the Sketcher. We will now learn to use the Sketcher. Let us first create a part on which to apply the Sketcher commands.

1. Start a new part by clicking the new icon ☐.

2. Leave **Solid**, **Part** and **Use default template** options *checked* in the New window.

3. Type *Sketched_Extrusion* as the name of file and click **OK**.

4. Click the **Extrusion Tool** 🗗 → **Sketch** 🖉 from the dashboard.

   (Version 2.0) Select Placement → Define...

5. In the **Placement** window, click the white space in front of **Plane**, as shown in Figure 3.1, to make it active, then select the **TOP** datum plane in the graphics window (Figure 3.2) and accept the default for **References** and **Orientation**.

6. Click **Sketch** Sketch .

7. The **References** window shows **fully placed with two references**. Close the References window.

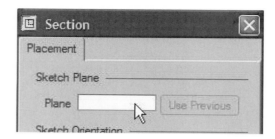

Figure 3.1

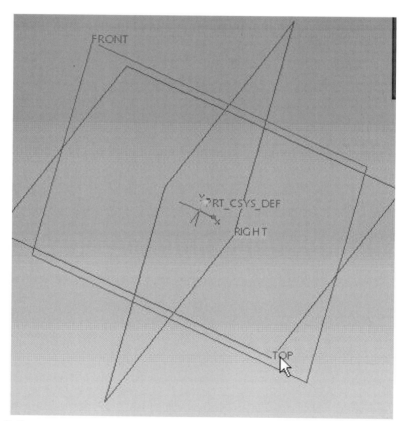

Figure 3.2

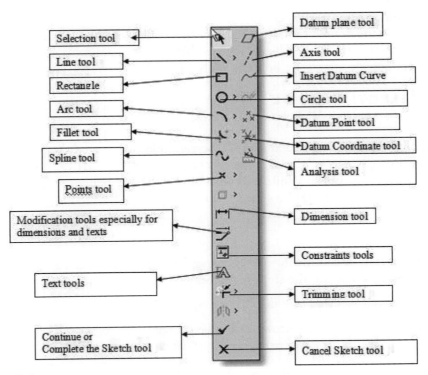

Figure 3.3

You are now in **Sketcher**. The sketcher tools, shown in Figure 3.3, are displayed to the right of the Graphics window.

Some of the tools have fly-out options, indicated by the arrows pointing to the right of the tools. The presence of the arrow indicates that more functionalities are hidden from view. Clicking the arrows reveals these additional functionalities. For example, clicking the right arrow of the line command brings up additional options as shown in Figure 3.4.

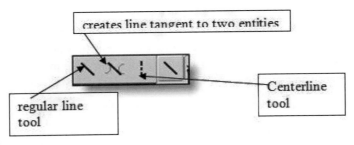

Figure 3.4

Information on other tools with fly-out options may similarly be obtained by resting the cursor on the desired tool. Let us demonstrate a few of these commands. Notice that the dimensions of the objects we are able to create are arbitrary, so that the reader can quickly appreciate how the tools are used.

THE SELECTION TOOL

We are representing a 3D solid on a 2D screen, selection can be tricky. As we will see later, Pro/ENGINEER makes the task easier by highlighting the object in focus as the cursor rests on it. While another sketcher tool is being used, clicking the middle mouse can activate the selection tool.

PRACTICE

Click the rectangle tool in the sketcher tool bar. Draw the rectangle as shown in Figure 3.5a by following the steps given below:

1. In the Sketcher, click the **Create rectangle** icon □.

2. On the graphics window, left-click in the upper left corner and drag the mouse cursor to the lower right corner and left-click again (Figure 3.5a).

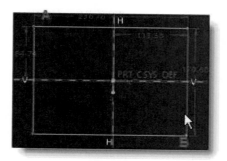

Figure 3.5a

3. Click the middle mouse once to end the rectangle command.

   Notice, on the features tool bar, that the rectangle tool becomes inactive and the selection tool becomes active. Clicking the middle mouse terminates the present command and then activates the **Select items** tool.

4. Adjust the dimensions so that the sketch looks like that shown in Figure 3.5b.

   A. With the Selection tool ▶ activated, draw a window around the sketched rectangle, so that *every* element of the sketch turns red.

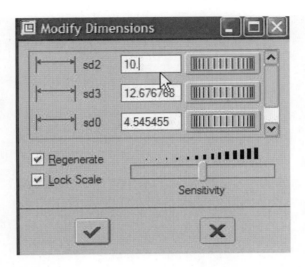

Figure 3.5b

B. Click **Modify Dimensions** tool and the **Modify Dimensions** window pops up. Check the **Lock Scale** option as shown in Figure 3.5c.

C. Change the vertical length of the sketched rectangle to **10** as shown in Figure 3.5c. Hit enter after keying in the 10. Then click the continue icon on the **Modify Dimensions** window.

D. Change the rest of the dimensions to those shown in Figure 3.5c by double left-clicking the dimension with the pointer tool. It is important that the location of your figure relative to the FRONT and RIGHT planes be identical to that shown in Figure 3.5c. If necessary, use ↶ to step back and repeat the Modify Dimensions operation.

E. If any dimension is missing, provide an additional dimension by clicking the **Dimensions** tool , clicking the two ends of the line and clicking the middle mouse wherever the text of the dimension is to be placed.

Note: The Modify Dimensions tool provides a way to scale a sketch proportionately.

THE LINE TOOL

The line tool is used for creating lines in the sketcher. In the previous exercise, we created a rectangle by using the rectangle tool. The same object could have been created using the line command.

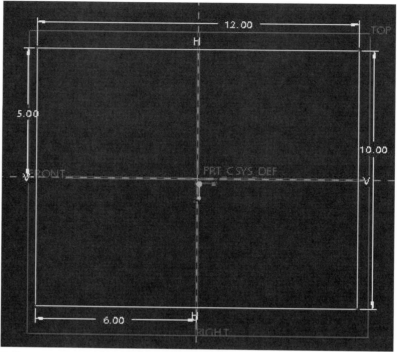

Figure 3.5c

1. Using the Sketcher tool bar, click the **Line Tool** .

2. Draw a diagonal to the rectangle (see Figure 3.6) followed by a middle mouse-click to terminate the line creation.

THE CIRCLE CREATION

The circle creation has an arrow pointing to the right (a fly-out arrow), indicating that additional tools options are available. The icon displayed in the list is the last option that was used. Other options of the circle, as shown in Figure 3.7, include the creation of concentric circles using , creation of circles using three points , creation of a circle tangent to three lines , and creation of an ellipse .

1. Using the Sketcher Tool bar, select circle option . If this circle option is not displayed, click the fly-out arrow to the right to reveal all the other options associated with the circle option.

2. Click at the intersection of the RIGHT and FRONT planes, to specify the center of the circle.

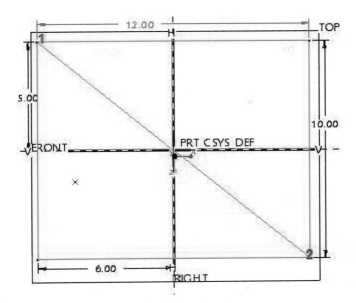

Figure 3.6

Figure 3.7

3. Drag the cursor outwards to vary the size of the circle. Click again when the circle attains the desired size (Figure 3.9). Middle-click the mouse to end the circle command. Middle-clicking again activates the Select Item tool ▸ in the Sketcher tool bar.

4. Double-click the dimension of the circle and change the diameter to **7**.

 Note: It is helpful to end a command and activate the Select Item tool bar to prevent accidental creation of unwanted sketches. Such errors will only become noticed when the designer attempts to quit the Sketcher environment. The system will often complain that the section is *incomplete*, if redundant sketches have been inadvertently created.

THE ARC CREATION

The arc creation has several options, as shown in Figure 3.8. Placing the cursor on any of the icons for a while reveals what these options represent.

Figure 3.8

1. Using the Sketcher tool bar, select .

2. Click on the diagonal line below the circle and at the intersection of the diagonal and the rectangle to the lower right of Figure 3.9. Zoom in with the middle mouse wheel, if necessary.

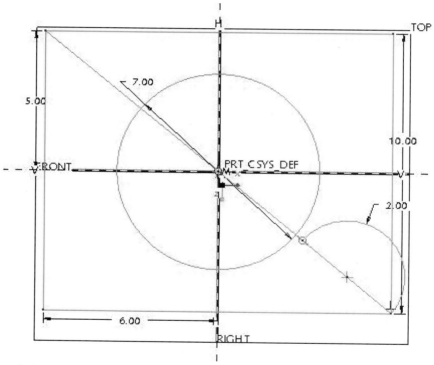

Figure 3.9

3. Drag the arc as shown in Figure 3.9 and click the left mouse to place the arc. (Be sure that the center of the arc lies on the diagonal of the rectangle.) Middle mouse-click to cancel the arc tool.

4. Double-click the *radius* and change the value to 2.0 (as shown in Figure 3.9). If the dimension is not visible, you can dimension an arc by left-clicking once on the arc and middle mouse-clicking at the desired location of the dimension text.

FILLET CREATION

The fillet creation has two options, as shown in Figure 3.10. Similarly, placing the cursor on the options and pausing for a while reveals descriptions of these options.

Figure 3.10

1. Select **Create Fillets** between two entities using the circular fillets  .

2. Click the left corners of the rectangle. Middle mouse to end the present command (the fillet command). Double-click the fillets and set their radii to **0.5** units to produce the fillets, as shown in Figure 3.11.

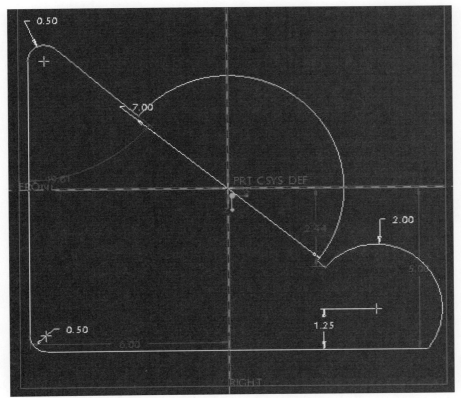

Figure 3.11

TRIMMING

Trimming is enabled by the options shown in Figure 3.12.

Figure 3.12

1. Click the **Dynamic trim** Tool .

2. Click the projecting ends of the top horizontal and right vertical lines to trim these off. Similarly, trim off the projecting end of the arc. Trim off the diagonal line between the ends of the bigger arc.

3. Zooming in by scrolling the middle wheel, if necessary, continue the trimming to create the section shown in Figure 3.13. (Zoom onto the lower right corner to clearly view the extensions being trimmed, if necessary.)

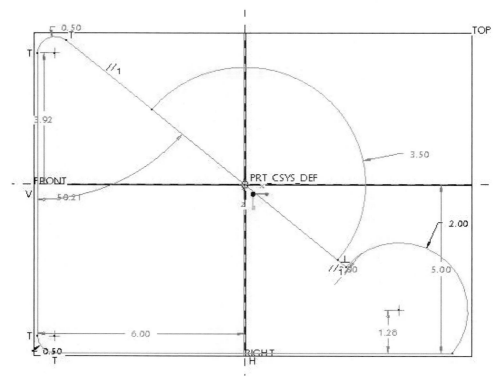

Figure 3.13

4. If an error is made during trimming, you can undo the operation by clicking the Undo tool in the standard tool bar ↶.

5. Middle-click to end the trimming operation.

COMPLETING THE SKETCH

Click on each of the weak dimensions and set them to the strong dimensions using the values shown in Figure 3.14.

Note: If a dimension has no weak specification, create the dimension by clicking the dimension tool ⊢⊣, clicking the two ends of the straight line and the middle mouse to place the value. In the case of the dimension of an arc, click the dimension tool ⊢⊣ and click once on the arc, followed by clicking the middle mouse at a location where the value will be placed. Clean up any overhanging sketch by trimming, if necessary.

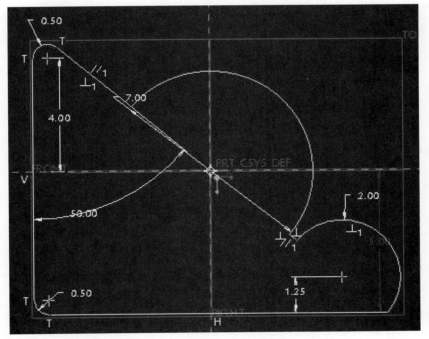

Figure 3.14

6. Select continue ✔ → ᴏᴋ (see Figure 3.15).

7. If an error occurs due to incomplete sketch, check if you have trimmed all loose ends. Are all the segments connected and continuous? Zoom-in on the ends of

Figure 3.15

the segments of the sketch and check that there is no break. If a break occurs, click 🔲 in the tool bar normally to the right-hand side. In the **Constraints** window that appears, click ⊙ →. Click the two broken lines to make them coincident. For example, the smaller arc may not have terminated on the diagonal line. In this case, you will click the end of the arc and then the end of the broken diagonal line to fix this problem. More examples on **Constraints** will be covered in the next section.

8. Set the extrusion height to **2.00**, as shown in Figure 3.16.

Figure 3.16

9. Click the Preview option 👓. Then view in the standard orientation by using Ctrl + D.

10. Toggle off the planes and coordinate system display using the standard tool bar. Your feature should be like that shown in Figure 3.17.

11. Click the Resume button ▶. Click Complete the feature ✔ to accept the feature as it is.

GEOMETRIC CONSTRAINTS

Recall Pro/ENGINEER's "sketch-and-adjust" philosophy we mentioned earlier. Now that we have sketched, the following are examples of constraints that may be imposed on our sketch to achieve a desired model:

1. Perpendicularity of lines ⊥

2. Symmetry of objects ⊣⊢

3. Point on entity ⊙

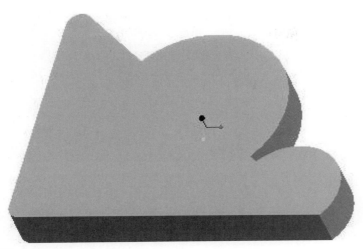

Figure 3.17

    4. Vertical constraints ↕

    5. Tangency ⌐

    6. Equality of sides =.

These constraints are only examples. Additional constraints exist and may be implemented intuitively as those presented here. The use of the geometric constraints help to achieve the intent of the designer. An example will make this clearer.

    1. Begin a new part ⬜.

    2. Name the part *constrained_shape* and accept the defaults in the New window.

    3. Click **OK**.

    4. Click the **Extrude Tool** 🗗 and then click the Sketcher Tool 🗹.

               (Version 2.0) Select Placement → | Define... |

    5. In the **Placement** window, click the white box in front of Plane to activate it. Click the TOP plane as the plane of sketch. The Sketch dialog box should look like that shown in Figure 3.18.

 Note: If the planes are not visible, be sure that the Datum plane Tool 🔲 in the standard tool bar is toggled on.

    6. Accept the default by clicking **Sketch** in the **Placement** window.

    7. Click **OK** to enter the Sketcher.

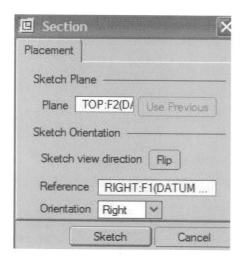

Figure 3.18

8. Using the line tool, draw the trapezoidal shape, shown in Figure 3.19, by left-clicking to specify the end of a line. Press the middle mouse to end line creation.

 Note: The dimensions are shown faint, because they are weak. As such, they are the dimensions based on the Pro/Engineers own internal computations about what these lengths are. Leave these dimensions free as they are. For simplicity, we are going to use free dimensions here and the dimensions indicated may be different from the dimensions displayed on your own screen. The object here is to get an experience of the process and not the specifications.

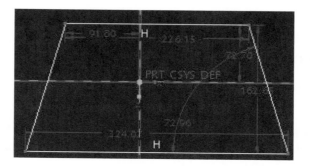

Figure 3.19

CONSTRAINT I

Make the vertical line perpendicular to the horizontal line at the left corner as explained in the following steps:

1. Using the Sketcher tool bar, click the **Constraints Tool** 🔲.

2. Select the **Perpendicular Tool** ⊥ in the Constraints window.

3. Select the two lines of the trapezium meeting at the lower left corner to make them meet at a right angle, as shown in Figure 3.20. Notice the appearance of the perpendicular symbol ⌐ close to the corner. Pro/Engineer Wildfire shows this symbol when two lines are perpendicular.

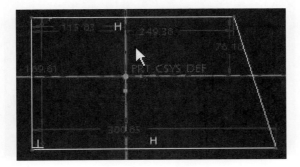

Figure 3.20

CONSTRAINT 2

Make the trapezium symmetric about the FRONT plane. For this we use another tool. However, *in order to create symmetry in Pro/ENGINEER, there must be an existing* **centerline.**

1. Click the arrow in front of the **Line Tool** ＼ to reveal more line options.

2. Click the **Centerline** option ＼✕┆＼. Then click any two points on the horizontal FRONT plane, away from the sketch itself, to create a centerline.

3. Click the **Sketcher Constraint Tool** 🔲. Select the **Symmetry Tool** ┿. Notice that the message on the dashboard asks you to select a centerline.

4. Select the centerline just created. Then, select the two points that will be symmetric about the centerline – these are the upper and lower tips of the vertical line to the left.

Pro/ENGINEER redraws this polygon to become as shown in Figure 3.21. Notice the *oppositely directed arrows* pointing towards the line of symmetry. This is the way Pro/Engineer indicates symmetry about a line.

If these arrows do not appear, symmetry has not been created. In that case, it will be necessary to repeat the process, remembering to first draw a centerline, before imposing the constraints. However, only one centerline is needed per sketch.

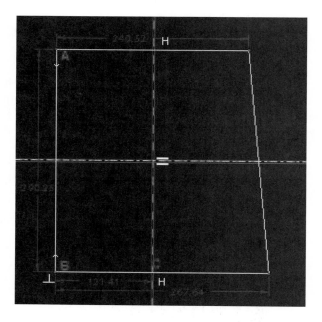

Figure 3.21

CONSTRAINT 3

Make the left side lie on the vertical (RIGHT) plane.

1. Select the Constraints Tool ▣.

2. Select the Create Same Point Tool ⊙ᵣ.

3. Select any point on the left vertical line AB in Figure 3.21 and then select any point on the RIGHT plane, such as point C in Figure 3.21.

Pro/ENGINEER makes the left side collinear with the RIGHT plane as shown in Figure 3.22.

 Note: If you attempt to make the lower side collinear with the horizontal (FRONT) plane, a conflict will arise. That is because, in one sense, you are saying the left bottom corner and the left top corner are symmetric about the FRONT plane and in another sense, you are saying that the left bottom also lies on the FRONT plane. Pro/E pops a window allowing you to undo the attempted change or delete one of the conflicting constraints.

CONSTRAINT 4

Make the sloping side vertical.

1. Select the Constraints Tool ▣.

2. Select the Vertical Tool ↕.

3. Select the right sloping side of the trapezium.

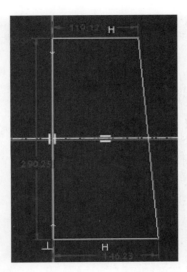

Figure 3.22

The sketch now appears as shown in Figure 3.23.

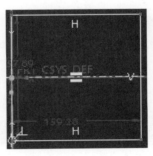

Figure 3.23

CONSTRAINT 5

Tangency

1. Create a circle of arbitrary radius to the right of the block previously created, so that your sketch is similar to that shown in Figure 3.24.

Hint: Click the Circle Tool ⬭, then click a point on the FRONT plane to the right of the existing sketch and then drag out the mouse to draw the circle with proportions similar to those shown in Figure 3.24 → Left-click to place the circle → Middle mouse-click to terminate circle creation.

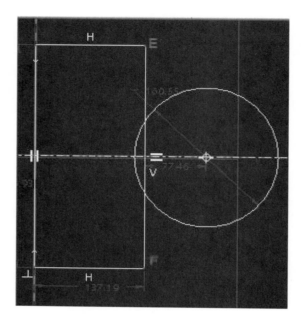

Figure 3.24

2. Click the Constraints Tool .

3. Click on the tangency constraint .

4. Click on the right vertical line of the rectangular block and then click on the circle.

The sketch now appears as shown in Figure 3.25.

CONSTRAINT 6

Make two lengths equal. We now wish to make the rectangular box into a square. We accomplish this by equating the length and the breadth of the rectangle.

1. Click on the Constraints Tool .

2. Click on the Equality Tool .

3. Click on the lengths and breadths of the rectangle.

Notice that the rectangle now becomes a square like that shown in Figure 3.26.

As it is above, we cannot create an extrusion – Pro/ENGINEER will warn that we have an incomplete section if we accepted the above as the completed sketch. In Pro/ENGINEER, extrusion requires a *continuous boundary* whereas the sketch above does not have a continuous boundary.

Complete the sketch by drawing two tangents to the circle as shown:

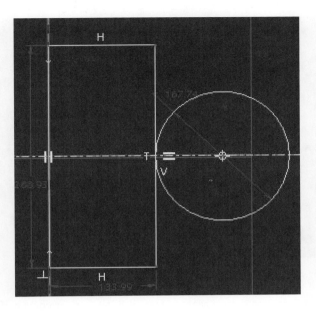

Figure 3.25

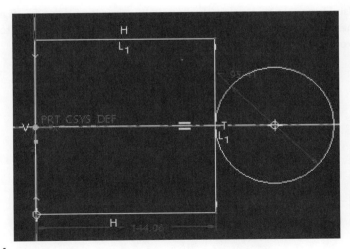

Figure 3.26

1. Close the Constraints window, if necessary.

2. Click the Line Tool ↘ → Select the upper right corner of the square and bring the cursor to rest on the circle until the tangent symbol appears, as shown in Figure 3.27 → Click the left mouse button → Click the middle mouse to terminate line creation.

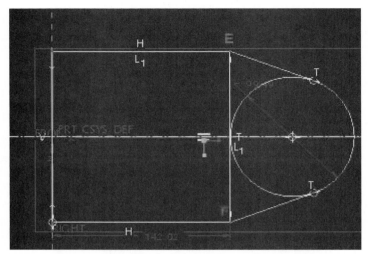

Figure 3.27

3. Repeat Step 2 to draw a tangent from the lower right corner of the block to the circle as shown in Figure 3.27.

4. Use **Dynamic trimming** ⚒, starting by trimming off line EF (Figure 3.27) and then trimming off the rest of the interior entities of the sketch to create a sketch similar to that shown in Figure 3.28.

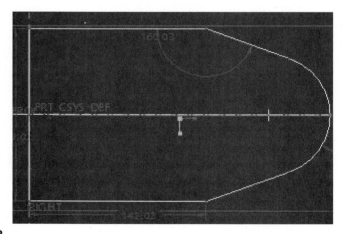

Figure 3.28

5. Click **Continue** ✓ → **OK** → ✓ (Dashboard) → Ctrl + D, so that the completed extrusion is similar to that shown in Figure 3.29.

Save the part and close its window.

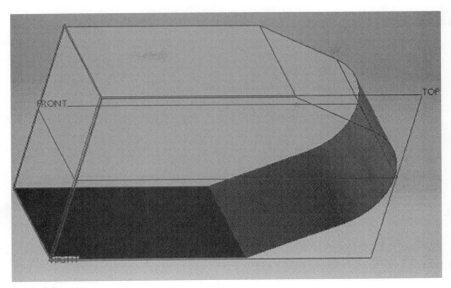

Figure 3.29

CREATING A POINT

The Datum Tool Point can be used to create a point in the sketcher environment or on a feature. The more useful tool is the **Offset Coordinate System Datum Point Tool**.

1. Start a new Pro/ENGINEER file and name it *datum_points*.

2. On the tool bar normally to the right-hand side of the graphics window, select . The **Offset Datum Point Tool** window appears. Notice that the dashboard now asks for the coordinate system. This is the coordinate system upon which the created point(s) will be based.

3. Select the **default coordinate system** .

4. In the **Offset Datum Point Tool** window, click under name and edit the name and the numeric values of the coordinates. Click in the space beneath name.

5. Observe that the created points are displayed on the graphics windows. If the points are not visible, verify that on the standard tool bar the **Datum Point** icon is toggled on.

EXERCISE

1. Create Figure Q3.1 and Figure Q3.2, ensuring that the constraints are satisfied. Save the files as sections files. To do this, click the save icon while in the Sketcher. (Saved section files can be reused.)

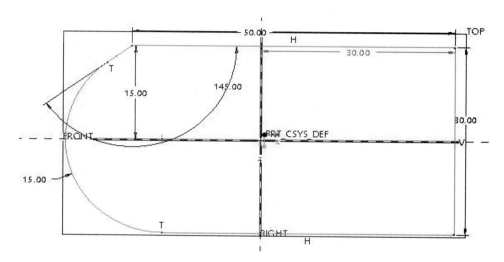

Figure Q3.1

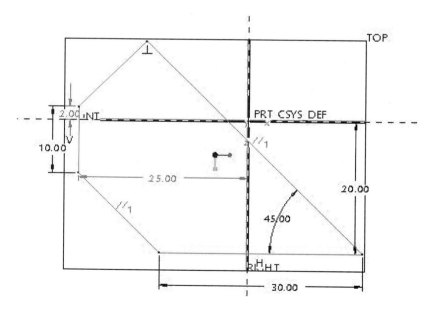

Figure Q3.2

CHAPTER 4

Revolved Solids

THE BASICS

Revolved solids have axial symmetry. Revolving a sketch to create a solid can be a very convenient method of creating solids with rounded surfaces. The following points are important regarding the creation of revolved solids:

1. Revolved solids are created by drawing a half cross section.

2. Every revolved solid must have an axis of revolution (created using the centerline tool in Sketcher).

We shall now create a revolved feature.

CREATING A REVOLVED FEATURE

1. Launch Pro/ENGINEER and set your working directory as appropriate (if needed).

2. Create a new part ⬜ and name it *Revolved_Solid*.

3. Accept the defaults in the New Part window.

4. Click the **Revolve Tool** in the features bar ⚙.

5. In the dashboard at the bottom of the screen, ensure that **Revolve as solid** ⬚ is toggled on. Your dashboard should then look like that shown in Figure 4.1. (Note that **Revolve as surface**, which we are *not* using yet, will produce a surface.)

6. Click the Sketcher Tool ✎.

(Version 2.0) Select Placement → Define...

Figure 4.1

7. In the **Placement** window that opens, click the white space in front of **Plane** followed by clicking the TOP plane in the graphics windows as the sketching plane.

8. Accept the defaults and click **Sketch** to enter the Sketcher. The **References** window shows **fully placed**. Click **Close** to begin sketching.

9. Select the centerline tool from the line option .

10. Click two different points on the vertical line, representing the RIGHT plane → Click the middle mouse to end this operation.

11. Select the Line Tool and using a series of left-clicks, create the sketch shown in Figure 4.2, ignoring the dimensions for now. Click the middle mouse to end the Line command.

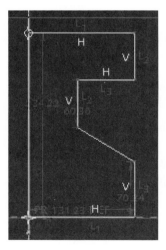

Figure 4.2

 Note: Also ensure that the sketch's left vertical line is actually drawn and that it rests on the vertical centerline previously created (in Step 10 above). If there was no left side of the sketch, our sketch would have been incomplete and Pro/ENGINEER will flag this when we attempt to continue ✔.

We are now ready to revolve the sketched feature about the centerline.

12. Click the **Complete the Sketch** button ✔.

13. View in the standard orientation Ctrl + D. Preview the feature by clicking 👓 in the dashboard.

14. In the displayed graphics, observe the position of the TOP plane relative to the created feature.

15. Click the **Resume Tool** ▶.

16. Toggle off the datum planes, datum axes, datum points, and coordinates, by clicking the tools 🔲 🖉 ✕ ✕ (if they are not already toggled off), for a clearer view.

17. Click **Complete the feature** ✔ to accept the created feature as it is. The created feature now looks like that shown in Figure 4.3.

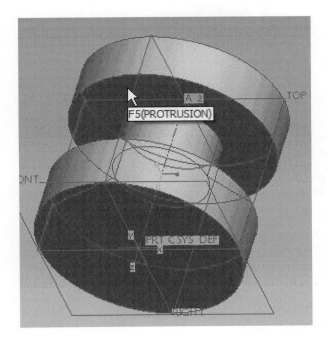

Figure 4.3

ADDING A REVOLVED SURFACE

We now wish to add a revolved surface on the revolved solid.

1. Toggle on the planes 🔲, if necessary.

2. By holding down the middle mouse in the Graphics window and simultaneously dragging it, position the created feature to look like that

shown in Figure 4.4. Hold down the shift key and drag to center the feature, if necessary.

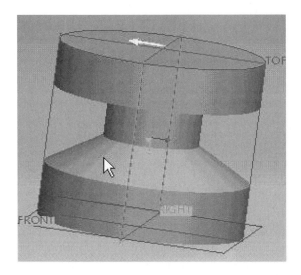

Figure 4.4

3. In the Features tool bar to the right of the Graphics window, click the **Revolve Tool** .

4. Toggle on the **Revolve as surface** Tool on the dashboard. Your dashboard should look like that shown in Figure 4.5.

Figure 4.5

5. Click **Sketch** .

(Version 2.0) Select Placement ⟶ Define...

6. The new feature is to appear on top of the part. In the Placement window, click the white space in front of **Plane** ⟶ In the Graphics window, select RIGHT plane as the sketching plane.

7. Again in the **Placement window**, click the textbox in front of **Reference**, then select the top surface of the feature to make it a reference.

8. For orientation, we want the top surface of the existing feature facing the top. In the textbox in front of **Orientation**, select top so that the orientation will be such that the top surface selected as reference is pointing to the top when displayed in the Graphics window.

9. Still in the **Placement** window, the **Reference** and **Orientation** settings should now look like that in Figure 4.6.

Reference	Surf:F5(PROTRUSI...
Orientation	Top ⌄

Figure 4.6

10. Click **Sketch** to enter the Sketcher.

11. The **Reference status** shows fully placed. Click **Close**.

12. Click **No Hidden** ⬜ in the standard tool bar for a clearer view.

13. The feature now looks like that shown in Figure 4.7.

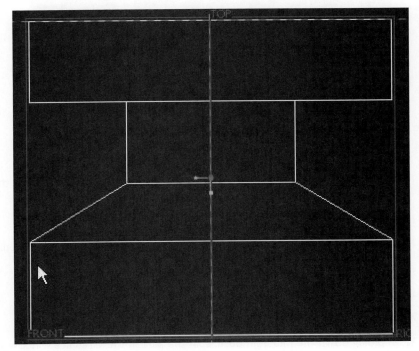

Figure 4.7

14. On the existing sketch, use the line command 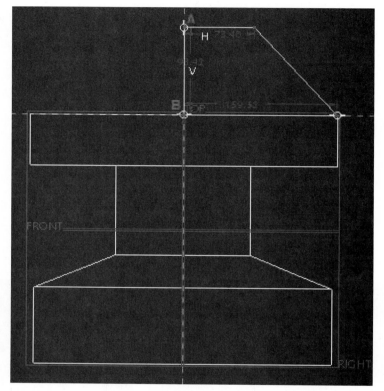 to create an additional sketch like that in Figure 4.8.

Figure 4.8

15. Add a centerline along the vertical side AB (Figure 4.8), by clicking ⁞ in the Line options and clicking any two points on the vertical line.

16. Select the **Continue** button ✔.

17. Click the **Shading Tool** ▢ on the standard tool bar to toggle back on.

18. Click the **Preview** button 👓 and the Resume button ▶.

19. Click **Complete the feature** icon ✔.

20. Toggle off the **Datum Planes** ▱.

21. Holding down the middle mouse, spin the part in the Graphics window so that it looks like that shown in Figure 4.9.

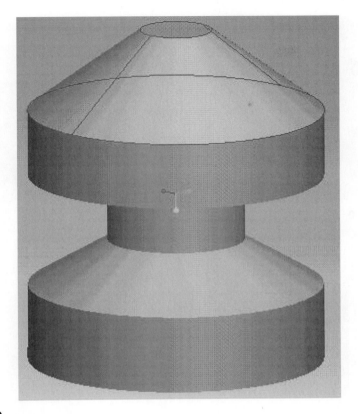

Figure 4.9

 Note: The display does not clearly show the new created feature to be a surface in Figure 4.9. That is because the surface is closed. We will now edit it, so that the display shows that a surface has been created.

EDITING A REVOLVED SURFACE

Let us edit the angle of revolution.

1. In the Model Tree to the left of the Graphics window, right-click on the Surface feature Surface id 114 (or Revolve feature in Version 2.0).

2. Select **Edit** in the pop-up window that is displayed.

3. Double-click on the angle of revolution shown on the entity in the Graphics (currently 360) window and *enter* **200**, as shown in Figure 4.10.

4. In Version 1.0, click **Regenerate Model** in the standard menu bar to see the effect of the change in the angle of revolution.

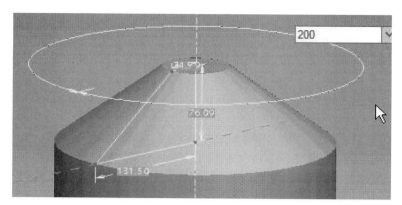

Figure 4.10

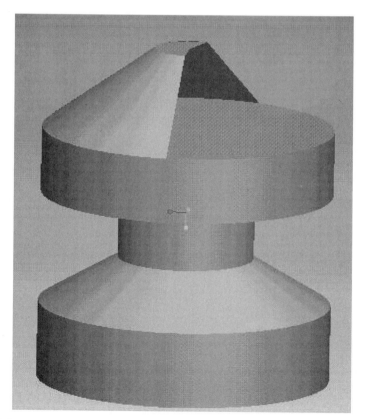

Figure 4.11

5. Using the middle mouse, spin the part so that it looks like that shown in Figure 4.11.

   The created entity can now be clearly seen to be a surface.

6. Accept and save the part.

EXERCISE

Create the following objects with the cross section shown.

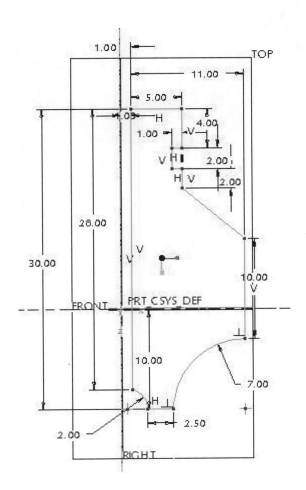

Figure Q4.1a

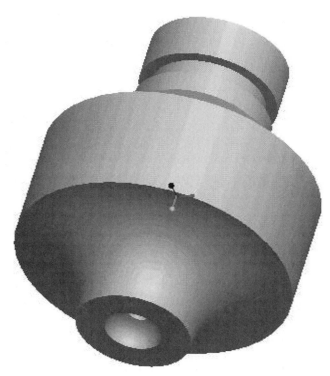

Figure Q4.1b

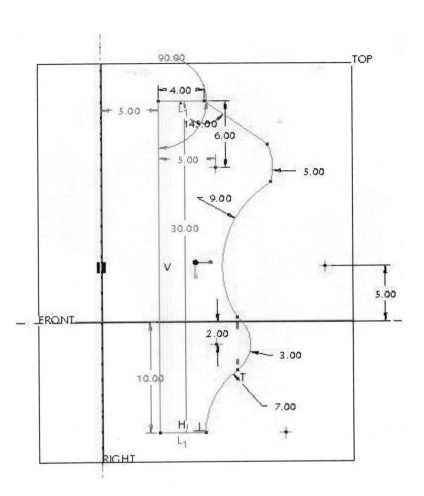

Figure Q4.2a

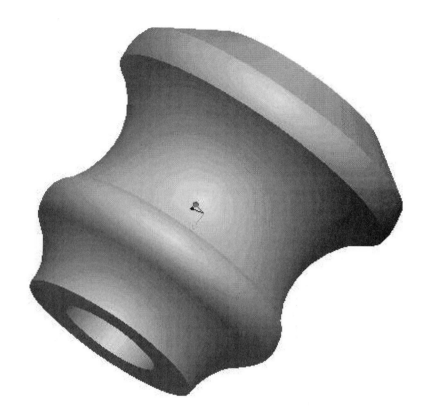

Figure Q4.2b

CHAPTER 5

Blends and Sweeps

THE BASICS

Blends, as the name implies, involves the joining of two sections by an interpolating straight line or curve. Sweeps, however, involve the moving of an entity around a prescribed path. Let us experience both by carrying out the following tasks.

CREATING A PARALLEL BLEND

1. Create a new part and name it *blends1*. Accept the defaults in the part creation window.

2. From the standard pull-down menu, click **Insert** → **Blend** → **Protrusion**.

3. Ensure that the options highlighted in Figure 5.1 are selected.

Figure 5.1

Note: The **Parallel** option means that the sections we are going to use in the blending operation are parallel to one another. **Regular Sec** option allows multiple sections to be blended and **Sketch Sec** means we are going to sketch the section, as opposed to using sections that may have previously been created and saved as a **Sketch** file.

4. Click **Done** → **Done**.

5. Ensure that the Datum planes icon [icon] is toggled on, if it is not already so. In the Graphics window, click TOP as the sketching plane. Click **Okay** → **Default** → **Close**.

 Note: Notice that the dialog window (Figure 5.2) shows that we are now Defining the Section (see the right arrow indication of the stage we are). The Pro/Engineer package is, in this way, helping the designer know the stage of the feature creation. We now draw the projections of the different sections of our part as viewed normal to the sketching (TOP) window. We can have a number of sections. However, the sections must match in the number of entities. **A point can be blended with any cross section**.

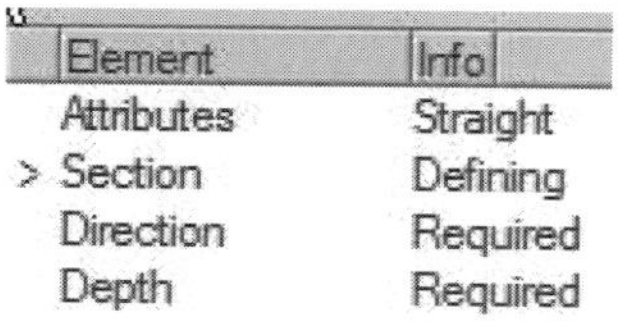

Element	Info
Attributes	Straight
> Section	Defining
Direction	Required
Depth	Required

Figure 5.2

6. Using the Rectangle Tool ▢ in the Sketcher tool bar, sketch a rectangle by clicking and dragging the mouse from the second quadrant to the fourth quadrant then left mouse-clicking, as shown in Figure 5.3. Middle-click to end the sketch.

7. Click the middle mouse button again to be sure that the **Select items Tool** ▸ in the Sketch tool bar is activated. Select the rectangle by dragging a window across it. Click the Modify Tool ▨.

8. In getting to Figure 5.3, it may be necessary to use the **Modify Dimensions Tool**. In the **Modify Dimensions** window, leave the **Lock Scale** as checked. Change the dimensions as shown in Figure 5.4 → Click [✔]. Modify the rest of the dimensions to correspond to those shown in Figure 5.3.

 Note: We have now completed the creation of *one* section of our part. We need to sketch another section to be blended with the section we have just created.

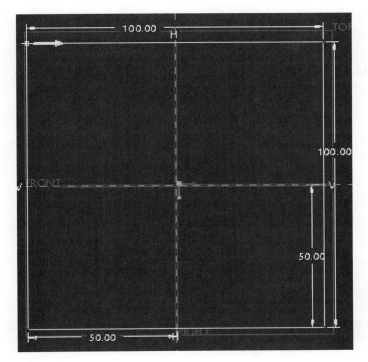

Figure 5.3

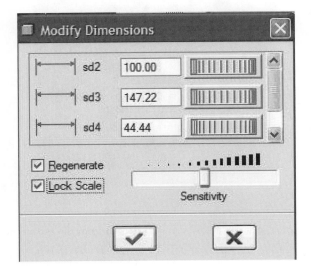

Figure 5.4

9. In the Graphics window, right-click and hold down for a while, until the options pop up and select **Toggle Section**. The previous section is grayed out, preparing us a new slate to create another section.

10. Using the Rectangle Tool □ and starting at the second quadrant, create another rectangle (Figure 5.5) enclosed within the first.

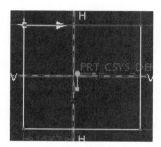

Figure 5.5

11. Click → Select the top horizontal line of the newly created rectangle → **Delete**. Draw an arc to modify the section as shown in Figure 5.6. Click the middle mouse to end the sketch creation.

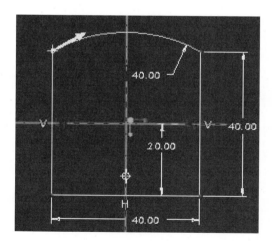

Figure 5.6

Note: Notice that the arrow sign has changed from the second quadrant to the first. The arrow sign denotes the starting point of each section. The blending operation matches the starting point of one section to another. The situation we now have is that the starting of one section is in the second quadrant while the starting point of another is in the first quadrant. If we accept the starting points as they are we will create a twisted section. We want to correct the starting point of the second section just created to also lie in the second quadrant.

12. Select the top corner of the left side of section 2, so that it turns red. Right-click and hold down → Starting Point. Change the dimensions so that the completed section now looks as in Figure 5.7.

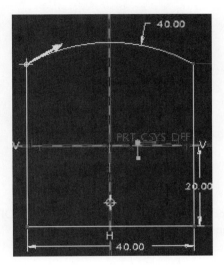

Figure 5.7

13. We now create a third section to be blended – a mere *point*! As before, in the Graphics window, right-click and hold down → Toggle Section. Select the Point Tool ✖. Click the origin (the intersection of the FRONT and RIGHT planes) → Click the middle mouse to end the sketch operation.

Note: We have created all sections as if they all lie on the TOP plane and as if they are parallel to one another. We will now be prompted to specify the distances between our created parallel sections.

14. Select the Continue button ✔. In the dashboard, enter **50** for the Depth of Section 2 (distance between sections 1 and 2). Similarly, enter **40** for the Depth of Section 3 (distance between sections 2 and 3). Recall that our Section 3 is a mere point, in this case.

15. Click **OK** from the menu manager in the upper right corner. Using the middle mouse, spin as appropriate to get a view similar to the one shown in the following figure. In Figure 5.8, notice that the sections are parallel to the TOP

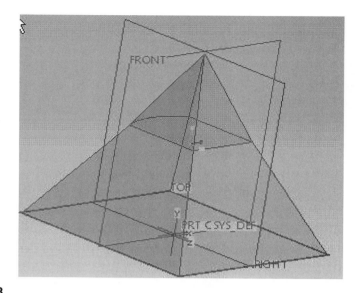

Figure 5.8

plane, as expected, since we had sketched on the TOP plane (or parallels to it, created by the off-setting distances).

16. Save the file.

CREATING A TWISTED BLEND

We now want to edit the *blends1* part created previously so that their starting points are offset.

1. With the *blends1* part opened, let us save a copy for the purpose of this task. Using the standard tool bar, click **File → Save a Copy**. Save the copy as *blends2*, as shown in Figure 5.9.

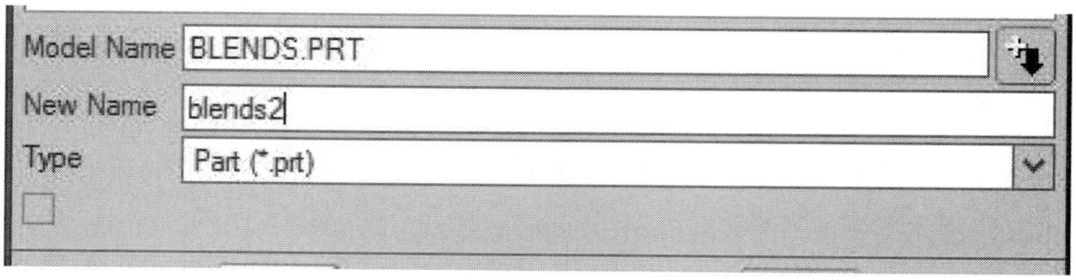

Model Name	BLENDS.PRT
New Name	blends2
Type	Part (*.prt)

Figure 5.9

2. Although we created a new file called *blends2*, it is not displayed. Hence, from the standard menu bar, close *blends* and open the newly created **blends2**.

3. Right-click the protrusion in the Model Tree and select **Edit Definition** as in Figure 5.10.

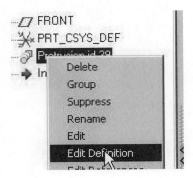

Figure 5.10

4. Select **Section**, as in Figure 5.11.

   Then click **Define** → **Sketch** → **Sketch** → **OK**.

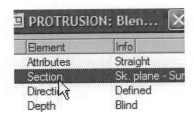

Figure 5.11

Note: The Sketcher opens up and the first section of the part is indicated.

5. Select the corner in the *first* quadrant, so that it turns red.

6. Right-click → **Start Point** → Continue ✔ → **OK**.

   The part now becomes as shown in Figure 5.12. Spin the part around as necessary to view from different sides.

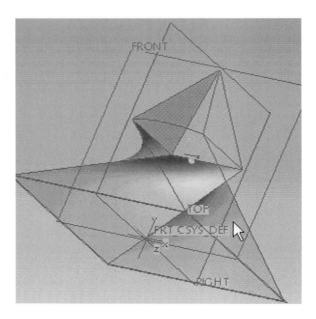

Figure 5.12

 Note: The main lesson here is that the designer must be deliberate about the choices he makes regarding starting points of blend sections.

7. Save and close the file (File → Close Window).

CREATING A SWEEP: CLOSED SECTION + OPEN TRAJECTORY SWEEP

Create a new part and name it *sweep1*. Accept the defaults in the part creation window.

1. From the standard menu, click **Insert** → **Sweep** → **Protrusion** → **Sketch Trajectory** → TOP (in the graphics window) → **Okay** → **Default** → **OK** → **Close**.

2. Create Spline by clicking ∿. Left-click at four different points, A, B, C, D, on the Graphics window and then middle mouse-click twice to create a spline similar to that shown in Figure 5.13. Place point A at the intersection of the FRONT and RIGHT planes. Use **Modify Dimensions** tool to scale appropriately.

3. Click the Continue Tool ✔. The Sketcher will reorient to be normal to the start point of the trajectory.

4. Note we have completed the trajectory creation.

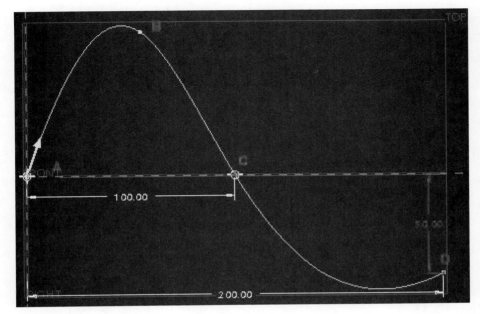

Figure 5.13

5. Sketch the section of the sweep by creating a rectangle centered at the intersection of the horizontal and vertical reference *golden* lines, as shown in Figure 5.14. Pro/ENGINEER has automatically figured this placement for you.

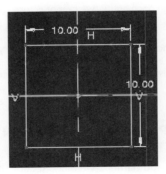

Figure 5.14

6. Click the Continue Tool ✔ → **OK**.

7. Click Preview. Press Ctrl + D to view the feature in standard orientation. Click OK.

8. Your completed view then looks like that shown in Figure 5.15.

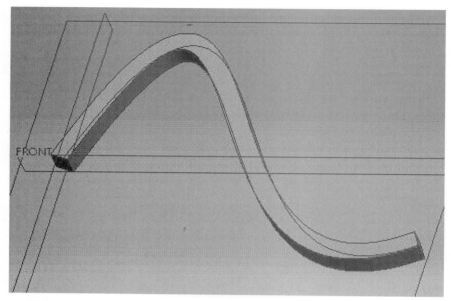

Figure 5.15

Note: It may be noted that what we have just accomplished is the pushing of our section (the rectangle) along a trajectory (the spline). Notice here that the section is a closed entity (we can traverse the entity and return to the beginning without overlapping our path) and that the trajectory is an open entity (we cannot return to the beginning of the entity without overlapping our path). There are other cases of moving an entity about another entity:

1. a closed section moved along a closed trajectory (no inner faces)

2. an open section moved along a closed trajectory (Pro/ENGINEER adds inner faces).

CREATING A SWEEP: CLOSED SECTION + CLOSED TRAJECTORY SWEEP

1. Create a new part and name it *sweep2*. Accept the defaults in the part creation window.

2. From the standard menu bar, click **Insert** → **Sweep** → **Protrusion** → **Sketch Trajectory** → TOP (in the graphics window) → **Okay** → **Default** → **OK** → **Close**.

3. Create the closed trajectory shown in Figure 5.16, starting at the arrow location and moving in the direction of the arrow. Notice that the semi-circle is

centered at the intersection of the RIGHT and FRONT planes and that it may be drawn using the tool 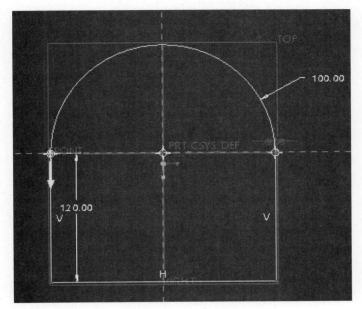.

Figure 5.16

4. Click the Continue tool ✓ → **No Inn Faces** → **Done**.

5. Using a combination of line, circle, and trim commands, create the *closed* section shown in Figure 5.17.

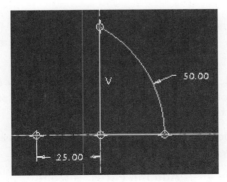

Figure 5.17

6. Click the Continue tool ✔ → **OK**.

7. Click Preview. Press Ctrl + D to view the feature in standard orientation. Click OK.

   Your section now looks like that shown in Figure 5.18.

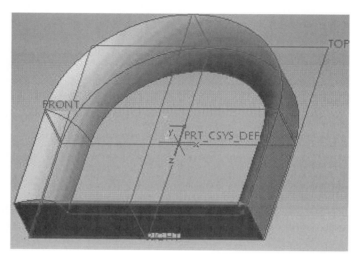

Figure 5.18

CREATING A SWEEP: OPEN SECTION + CLOSED TRAJECTORY SWEEP

1. Create a new part and name it *sweep3*. Accept the defaults in the part creation window.

2. From the standard menu, click **Insert → Sweep → Protrusion → Sketch Trajectory → TOP** (in the graphics window) → **Okay → Default → OK → Close**.

3. Create the closed trajectory shown in Figure 5.19 by clicking ▢ and dragging from the second quadrant to the fourth quadrant followed by a middle mouse-click to end the command.

4. Click the middle mouse button → select each dimension in turn and set the values to correspond to those shown in Figure 5.19.

5. Click continue ✔ → **Add Inn Faces → Done**.

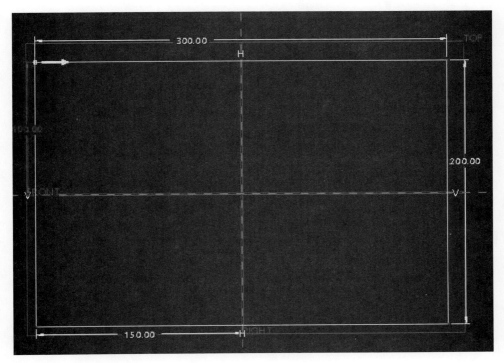

Figure 5.19

6. Using the arc command , create the section consisting *only of the arc* of radius **100** units, as shown in Figure 5.20. Notice that your arc ends must lie on the reference lines and be centered on their intersection. The reference lines are indicated as dotted golden color lines on the Graphics window.

7. Click the Continue tool ✔ → **OK**.

8. Click Preview. Press Ctrl + D to view the feature in standard orientation. Click OK.

Your feature now looks like that shown in Figure 5.21.

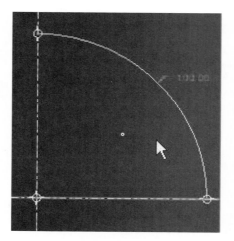

Figure 5.20

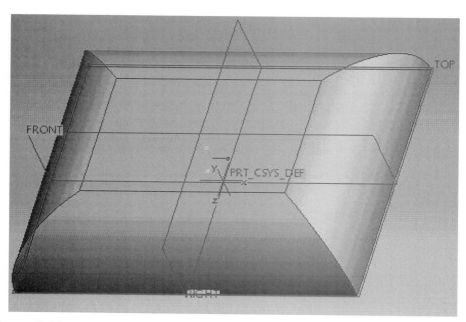

Figure 5.21

EXERCISE

1. Right-click the protrusion of *Sweep3* in the Model Tree. Select **Edit Definition**. Modify the section by adding a few more entities to the arc so that the section becomes like that shown in Figure Q5.1.

2. Click the Continue Tool ✔ → **Done**. Click Preview. Click **OK**.

3. Use the middle mouse wheel to adjust the view to be like that shown in Figure Q5.2

4. Using free dimension, create the part shown in Figure Q5.3.

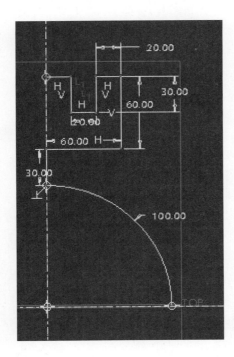

Figure Q5.1

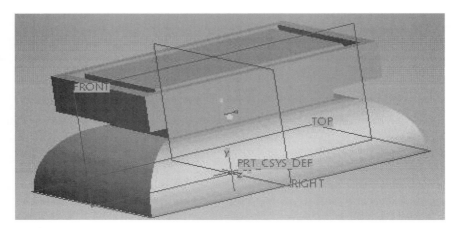

Figure Q5.2

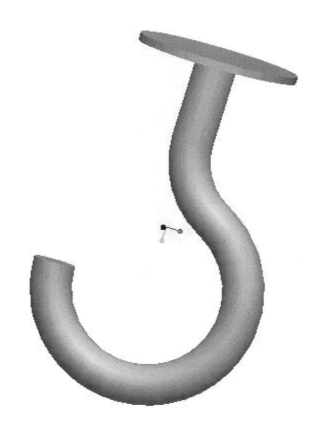

Figure Q5.3

CHAPTER 6

Drawing

THE BASICS

Drawings provide 2D representations of a model. A 2D representation of a model can easily be generated from the part model. Three-dimensional representation of solid objects will seem to appear more natural. Pro/Engineer enables the designer to create a part as a 3D entity from which a 2D drawing may automatically be generated. We shall now create a simple 3D part and generate its corresponding 2D drawings.

CREATING A RECTANGULAR BASE

1. Start a new file ▢.

2. In the **New** window, select the Type as Part. Name the part as *drawing_part*.

3. Click **OK**.

4. Using the menu bar, select **Insert** → **Extrude**.

5. Click the Sketcher tool to go into the sketcher ▱.

   (Version 2.0) Select Placement → ⬚ Define...

6. Select the TOP plane in the Graphics window as the sketching plane. Click **Sketch** → **Close**.

7. Sketch the entity shown in Figure 6.1.

8. Click continue ✔.

9. In the dashboard, set the depth to **20** and the rest of the dashboard setting should be as shown in Figure 6.2.

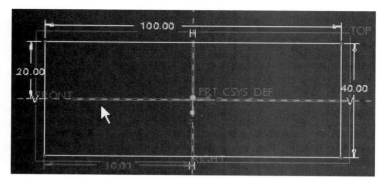

Figure 6.1

Figure 6.2

10. Click Complete the feature ✔.

11. Click Ctrl + D to view in the standard orientation. Your completed part should be like that shown in Figure 6.3.

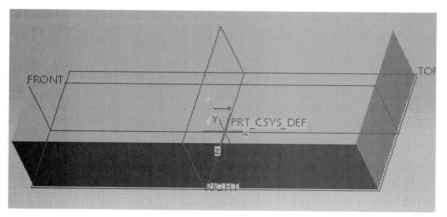

Figure 6.3

We are now going to put a cylinder on the top of the block. We should therefore select the top surface as our sketching plane when prompted. We will create the cylinder as another extruded protrusion.

CREATING THE CYLINDRICAL COMPONENT

1. Using the menu bar, select **Insert → Extrude**.

2. Click the Sketcher tool to go into the sketcher ☑.

   (Version 2.0) Select Placement → Define...

3. Select the top of the box in the Graphics window (anywhere on plane ABCD in Figure 6.4, such as that indicated by the cursor position) as the sketching plane.

4. Set the orientation such that the FRONT plane – presently facing to the back – is facing the top, as shown in Figure 6.5.

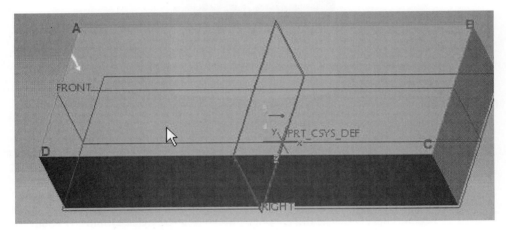

Figure 6.4

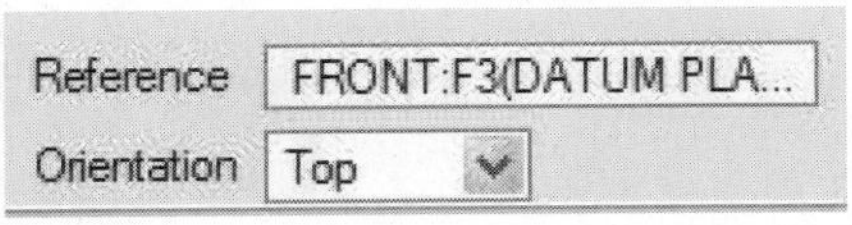

Reference	FRONT:F3(DATUM PLA...
Orientation	Top

Figure 6.5

5. Click **Sketch → Close**.

6. Toggle on the **No Hidden** tool for a clearer view 🗇.

7. Using the Circle Tool ⭕, sketch a circle centered at the intersection of the FRONT and RIGHT planes. (⭕ → Click intersection → Drag mouse outwards → Middle mouse-click to end circle creation.) If prompted to align the objects, accept Yes.

8. Click the Selection Tool 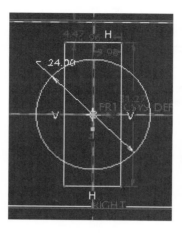 → Double-click the weak diameter value → Set the diameter of the circle to **24**.

9. Draw a rectangle across the circle, as shown in Figure 6.6.

Figure 6.6

10. Then use the Dynamic trim Tool to modify the sketch so that it becomes like that shown in Figure 6.7a.

11. Click the Selection tool → Select each dimension → Set the dimensions as shown in Figure 6.7a. Impose symmetry using → → Create centerlines → Select the endpoints of a line to be made symmetric about the

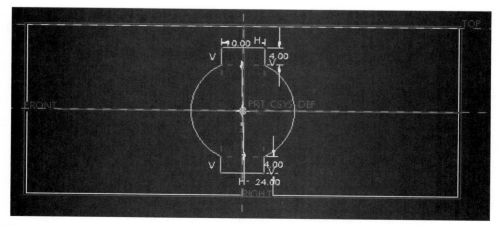

Figure 6.7a

respective centerline. Impose horizontality constraint using  → Click ↔ to make the ends of the lines have a level height, as appropriate.

12. Click continue ✔.

13. In the dashboard, set the depth to **20** and the rest of the dashboard setting should be as shown in Figure 6.7b.

Figure 6.7b

14. Click Complete the feature ✔.

15. Click Ctrl + D to view in the standard orientation. Toggle on the **Shading mode** ▢. Your completed part should be like that shown in Figure 6.8.

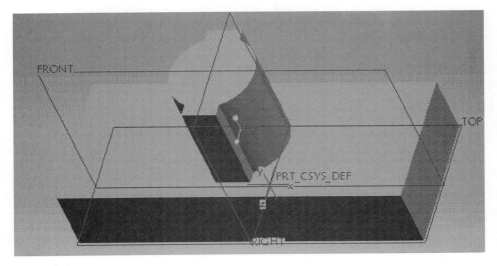

Figure 6.8

GENERATING THE ENGINEERING VIEWS

Now, let the fun begin. We are going to automatically generate the engineering views of the above part.

1. With *DRAWING_PART* part opened, start a new file ▢. In the New window, select the type as **Drawing** → Type *drawing1* as the file name, as shown in Figure 6.9.

2. Accept the default template and click **OK**.

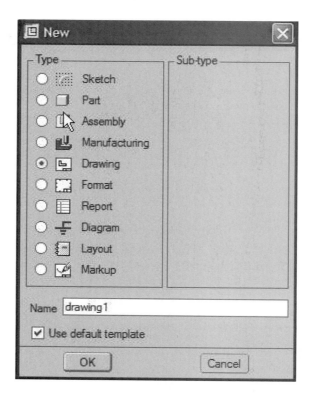

Figure 6.9

3. In the New Drawing window that opens up, ensure that *DRAWING_PART* is selected as the model. Set the New Drawing window as shown in Figure 6.10.

Note: Whatever drawing is displayed in the Graphics window is used by Pro/ENGINEER as the default model. If none is displayed, a more recent file may be displayed.

Notice that we are using the "c-drawing" template (a Pro/ENGINEER default). At your time, experiment with other types of templates.

4. Click **OK**.

5. Pro/ENGINEER automatically generates the engineering drawings of our part as shown in Figure 6.11.

6. The relative size of these views can be changed by double-clicking on the **Scale** value in the bottom left corner and entering a preferred value.

We are now going to insert an additional view to the drawing.

1. In the standard tool bar, click **Insert Drawing View ...** (or click the **Create a General View** icon).

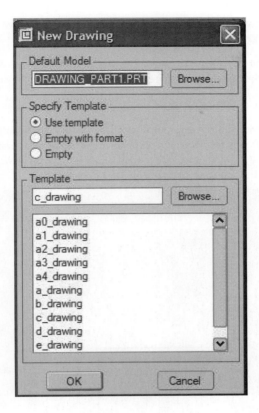

Figure 6.10

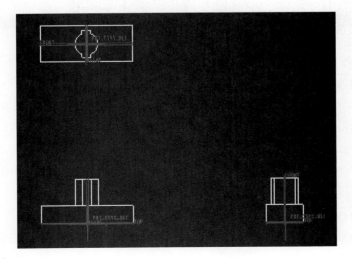

Figure 6.11

2. In the **View Type** menu, select **General** → **Full View** → **No Xsec** → **No Scale** → **Done**.

3. Click in the top right corner of the graphics window → **OK**, to place an auxiliary view of the part. The displayed views are now similar to those shown in Figure 6.12.

 Note: Version 2.0 simplifies the above three steps by presenting the user a form, or a dialog box, in which similar information as above is entered. Version 2.0 also requires the user to specify center point before displaying the Drawing view.

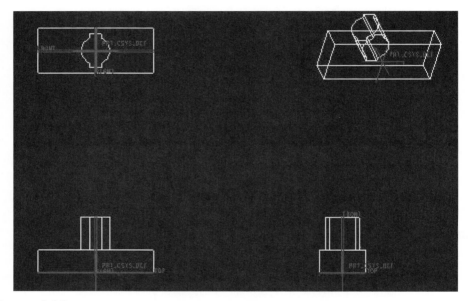

Figure 6.12

MOVING THE VIEWS

The views can be moved by toggling on the icon 🖥️ in the standard tool bar.

1. If necessary, toggle on the Movement Tool 🖥️.

2. Click each view, so that the entire view turns red and the movement icon 🔶 shows up → drag the views closer to themselves, so that the arrangement resembles that shown in Figure 6.13.

DIMENSIONING THE VIEWS

1. In the standard tool bar, click the Show and Erase icon ▧ to toggle it on.

The Show/Erase window opens up, portion of which is shown in Figure 6.14.

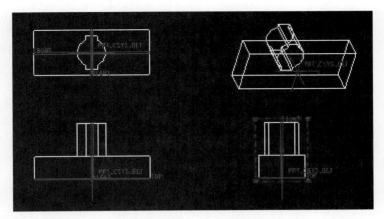

Figure 6.13

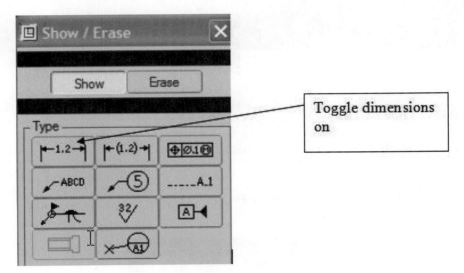

Figure 6.14

2. In the Show/Erase window, click **Dimension** as shown in Figure 6.14 → click **Show All** → **Yes** → **Accept All** → **Close**.

3. Use the middle mouse to zoom in, if necessary, to see the values clearly.

4. Select each dimension of a view, as appropriate, so that it turns red. Drag the dimension and leader lines to a desired location, for clarity. You can also right-click on each dimension and select flip arrow for smaller dimensions. The dimensions themselves can be selected and dragged as appropriate. Your final drawing should be like those shown in Figure 6.15.

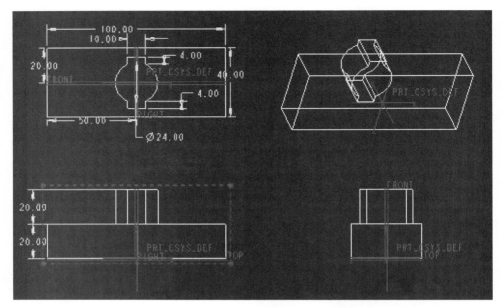

Figure 6.15

INSERTING A NOTE

1. Using the standard menu bar, click **Insert → Note → Leader → Enter →
   Horizontal → Standard → Default → Make Note**.

2. Select the top edge of the model as shown by the cursor position in Figure 6.16.

3. **Select On Entity → Arrow Head → Done**.

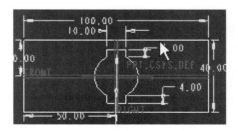

Figure 6.16

4. In the Graphics window, select a location for the note, as shown in Figure 6.17.

5. At the bottom of the screen, type *BACK FACING*. Click ✓ → Click
   ✓ → **Done/Return**. An annotation such as that shown in Figure 6.17 is created.

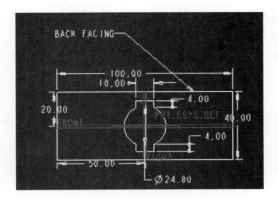

Figure 6.17

CHANGING THE FONT

1. Using the standard menu bar, click **Format** → **Text Style**.

2. Select the text BACK FACING, if it is not already selected, so that it turns red.

   The **Text Style** window, shown in Figure 6.18, pops up.

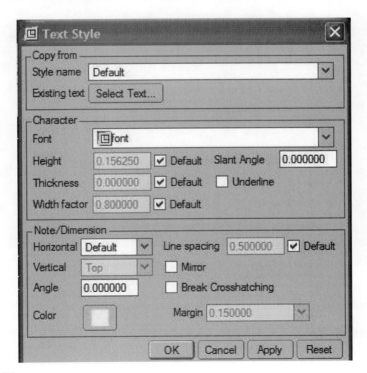

Figure 6.18

3. Uncheck the boxes on the sides of the **Height** and **Thickness** in the **Text Style** window.

4. Set the **Height** to **0.4**, the **Thickness** to **0.02**, and the **Slant Angle** to **30**, as in Figure 6.19.

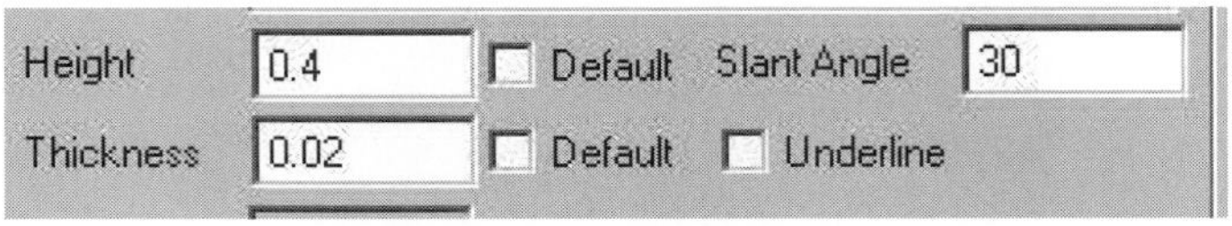

| Height | 0.4 | ☐ Default | Slant Angle | 30 |
| Thickness | 0.02 | ☐ Default | ☐ Underline | |

Figure 6.19

5. Click **OK**.

The drawing now looks like that shown in Figure 6.20.

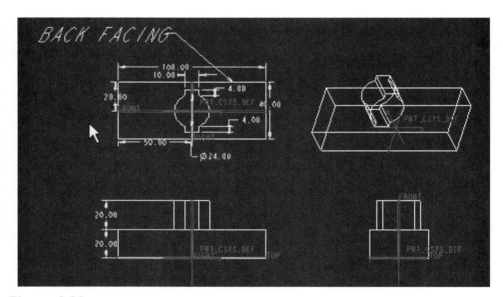

Figure 6.20

6. Select and drag the text around the drawing, as desired.

CHANGING THE LEADER PROPERTIES

1. From the standard menu, click **File** → **Properties** → **Drawing options**.

2. The Options window, shown in Figure 6.21, pops up.

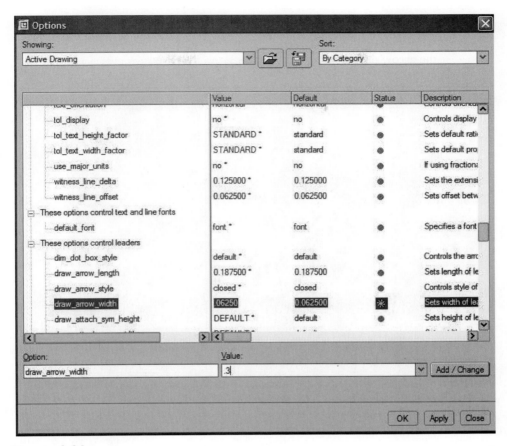

Figure 6.21

3. Scroll down the **Options** window until you find the section: "These options control leaders".

4. Select the draw_arrow_width. Change the value to **0.03**, as shown in Figure 6.22 → Click **Add/Change**.

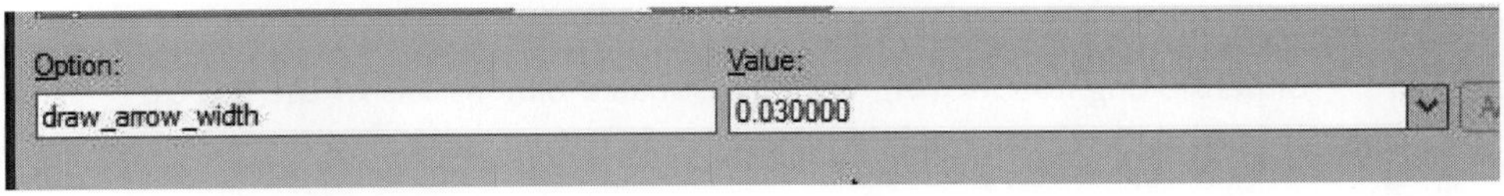

Figure 6.22

5. Similarly, change the **draw_arrow_length** to **0.1**. Click **Add/ Change** → **Apply** → **Close** → **Done/Return**.

6. In the drawing window, click the Update Model icon ⊞↻.

Notice that the arrow dimensions have changed, as shown in Figure 6.23. Reset the arrow properties to other dimensions, if desired.

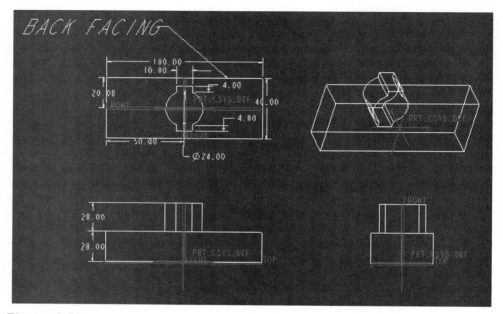

Figure 6.23

DRAWING A TITLE BOX

Most people will create a format that may be reused for their title box. What is shown here is a "quick-and-dirty" way to insert a title box in Pro/ENGINEER.

1. Rearrange the views, if necessary, to create sufficient space at the lower right corner for the title box.

2. Click the Create Line Tool in the drawing window ⟍.

3. Draw the boxes, as shown in Figure 6.24, to represent the title boxes (dimensioning not shown), using reasonable dimensions of the boxes.

Note: As you will quickly find out, in the drawing environment, the line command only creates a single line between two clicks and a single line between two successive clicks. We now use Insert Note to enter the indicated texts in the title box.

4. Select **Insert → Note → No Leader → Enter → Horizontal → Standard → Default → Make Note → Pick Pnt**.

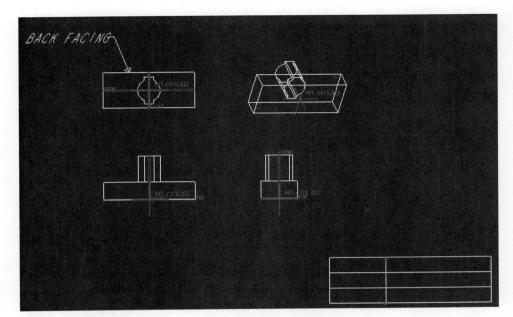

Figure 6.24

5. Select an appropriate location for the note within the cell (1,1).

6. Type *Designed By* → Click ✓ → Click ✓ → **Make Note**.

7. Repeat the steps for the other cells and enter the corresponding texts.

8. Click **Done/Return** to end the note insertion.

9. Click each text until it turns red, then reposition within the box as appropriate, so that the drawing now looks like that shown in Figure 6.25.

INSERTING A SECTIONAL VIEW

Let us now insert a sectional view of our part.

VERSION 1.0

1. In the standard tool bar, click Insert Drawing View.

2. In the View Type menu, select **Projection** → **Full View** → **Section** → **No Scale** → **Done**.

3. In the XSec Type, select **Full** → **Total Xsec** → **Done**.

4. Click above the upper left view to indicate to Pro/ENGINEER where to place the section → **OK**.

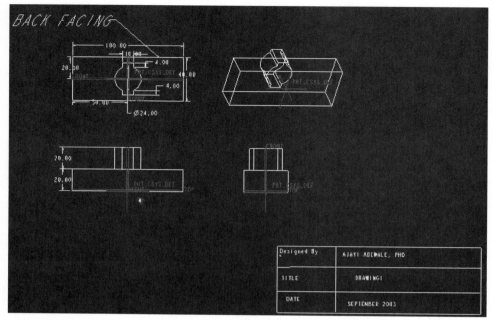

Figure 6.25

5. In the Xsec Center menu, select **Create** → **Planar** → **Single** → **Done**.

6. Enter R as name of the section. Entering R as section name means when the section is created, it will actually be named R-R.

7. Click the FRONT plane to indicate to Pro/ENGINEER the sectioning plane.

8. Click in the top view again to indicate the view to place the sectioning arrows (as shown in Figure 6.26).

Note: If an error occurs, you can delete the section and repeat the process described above. Simply select the section, hit the delete key and refresh the view. Refreshing a view to update display is no longer necessary in Version 2.0 – as part of the upgrade of this release.

VERSION 2.0

A variation exists for Pro/ENGINEER Version 2.0 in the sense that a dialog box acquires the above information from you.

1. Select the lower left view → Right-click → **Properties**.

2. Select **Sections** → Click the **2D cross section** → Add Section + → **Create New** → **Planar** → **Single** → **Done**.

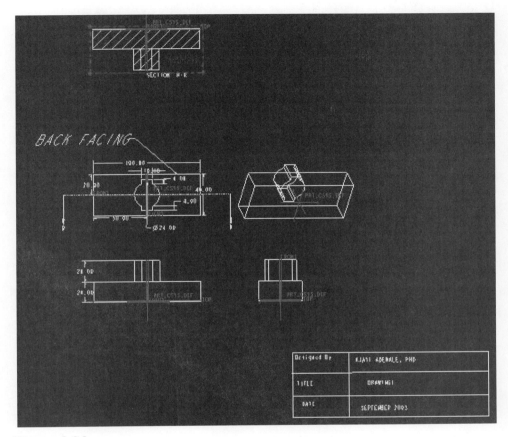

Figure 6.26

3. Enter *V2* as the section name.

4. **Plane** → Select the FRONT plane → **OK**.

5. Select the view, if it is not already selected → Right-click → **Add Arrows** → Click the top left view.

6. Notice that the section has *replaced* the view. If the view was desired, then using the menu bar select **Insert** → **Drawing View** → **Projection** to create a view.

ADDING DETAILS
VERSION 1.0

Let us now insert a detailed view of our part. The insertion of interest now is a *magnified* view of a portion of a part. Magnification is achieved by using the **Scale** option.

1. In the standard tool bar, click Insert Drawing View ⊞.

2. In the View Type menu, select **Detailed** → **Full View** → **No Xsec** → **Scale** → **Done**.

3. Click to the right of the sectioned view (Section R-R) to indicate to Pro/ ENGINEER where to place the detailed view.

4. In the scale prompt, enter **0.05** (at the bottom of the screen).

5. Pick the corner located at the junction of the horizontal and vertical bars of the section, as shown in Figure 6.27, as the center of the detail.

Figure 6.27

6. Click *around* the intersection picked in the above step, to sketch a spline around the intersection. Middle-click to end the drawing of the enclosing spline.

7. Enter A as the name of the view (at the bottom of the screen).

8. Select circle as the boundary type.

9. Click the location where the detailed note will be placed.

The completed drawing then looks as shown in Figure 6.28.

VERSION 2.0

1. Ensure that none of the views is selected by clicking in the background.

2. Using the standard menu bar, click **Insert** → **Drawing View** → **Detailed**.

3. Zoom in and pick the corner located at the junction of the horizontal and vertical bars of the section, as shown in Figure 6.27, as the center of the detail.

4. Sketch a spline around the corner and middle-click to end spline creation.

5. Click a location to place the detail.

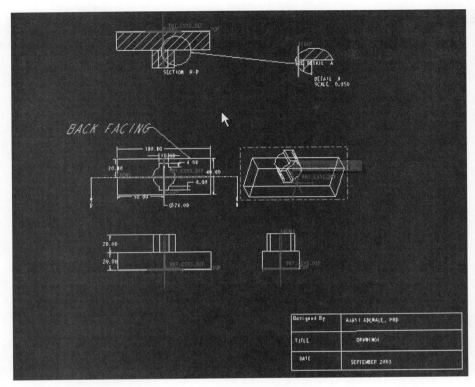

Figure 6.28

EXERCISE

1. Create the part in Figure Q6.1a and Figure Q6.1b, showing the views. Include the title box.

2. Create the part in Figure Q6.2a and Figure Q6.2b, showing the views. Include the title box.

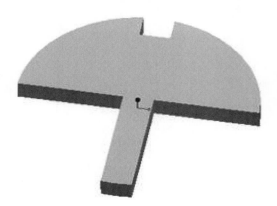

Figure Q6.1a

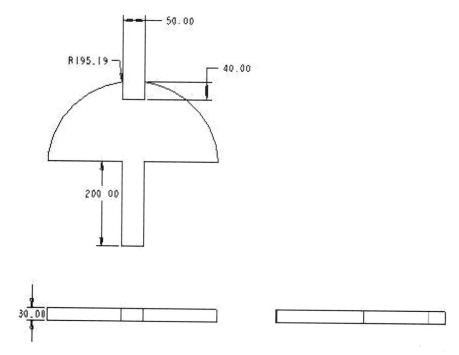

Figure Q6.1b

Figure Q6.2a

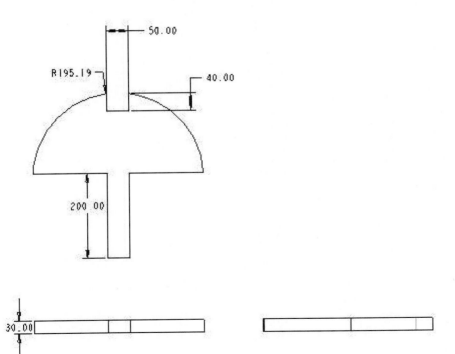

Figure Q6.2b

CHAPTER 7

Assembly

We now come to another fun part: Assembly mode. An assembly is just what the name implies: a collection of several components. Assembly in Pro/ENGINEER is accomplished through the use of constraints. Common constraints are mate, align, insert, default, and coordinate.

THE MATE CONSTRAINT

The mate constraint brings the two surfaces together, *facing each other*. There are three types:

1. Mate coincident (or simply mate) leaves no clearance. This is a case of *intimate contact* of the mating surfaces.

2. Mate offset leaves an offset distance between the surfaces being brought together (user supplies distance).

3. Mate orient.

THE ALIGN CONSTRAINT

The align constraints cause two entities to point in the *same direction or become parallel*. For axes, the realization is quite obvious. For surfaces, their normals are caused to face in the same direction. There are three types:

1. Align offset causes planar surfaces to become parallel but requires that an offset distance between the plane's edges be specified. An aligned offset therefore implies a *specified offset*.

2. Align coincident (or simply align) simply causes the surfaces or axis to face the same direction. Coincident here means that the edges line up. It is a case of *zero offset*.

3. Align orient causes planar surfaces to become parallel, similar to the align offset, except that the offset distance specification is not specified. Hence align orient allows the user to still freely move one entity relative to another in an orientation direction. It is a case of *variable offset*.

THE INSERT CONSTRAINT

The constraint is self-explanatory. However, it is limited to objects with axis. It is used in placing our axisymmetric object inside another axisymmetric object. For example, it may be used for placing a bolt in a hole. This constraint aligns the two axes of symmetry of the male and female objects.

THE COORDINATES CONSTRAINT

This can be used to precisely locate one object relative to another. It requires the creation of two coordinate systems and co-locating the two coordinate systems. Let us now demonstrate with some examples. But we shall first build the parts, which we wish to assemble. The parts consist of a bushing, a cylinder, a box, and a cover.

EXAMPLE
CREATING THE BUSHING

1. In the Pro/ENGINEER window, start a new part and name it *asm_part1*. Accept the defaults.

2. From the standard menu bar, click **Insert → Extrude**.

3. In the dashboard at the bottom of the screen, click the Sketcher tool.

   (Version 2.0) Select Placement → Define...

4. In the Graphics window, click the TOP plane as the sketching plane and click **Sketch** to accept the defaults. Then click **Close** in the reference box.

5. Using the circle tool ⃝, sketch a circle centered at the intersection of the FRONT and RIGHT planes. (⃝ → Click intersection → Drag mouse outwards → Left mouse-click → Middle mouse-click to end circle creation.)

6. Repeat the above step for a smaller circle, concentric with the previous one.

7. Middle mouse-click to enable the selection tool, or click the selection tool from the tool bar.

8. Click the diameter dimensions and set them to 4 and 8, respectively.

9. Click continue ✔.

10. In the dashboard, set the height to **2.00** (see Figure 7.1).

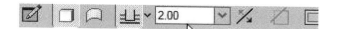

Figure 7.1

11. Complete the feature ✔.

12. View in the standard orientation using Ctrl + D.

    The completed part now looks as shown in Figure 7.2.

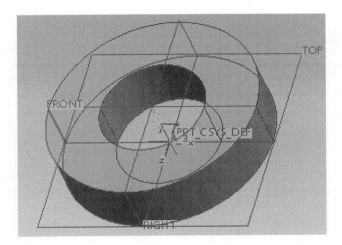

Figure 7.2

13. Save the file.

CREATING THE CYLINDER

1. In the Pro/ENGINEER window, start a new part and name it *asm_part2*. Accept the defaults.

2. From the file menu, click **Insert** → **Extrude**.

3. In the dashboard at the bottom of the screen, click the Sketcher Tool ▧.

    (Version 2.0) Select `Placement` → `Define...`

4. In the Graphics window, click the TOP plane as the sketching plane and click **Sketch** to accept the defaults. Then click **Close** in the **Reference** box.

5. Using the circle tool ◯, sketch a circle centered at the intersection of the FRONT and RIGHT planes. (◯ → Click intersection → Drag mouse outwards → Left mouse-click → Middle mouse-click to end circle creation.)

6. Middle mouse-click to enable the selection tool ⬚, or click the selection tool ⬚ from the tool bar.

7. Click the diameter dimension and set it to **4**.

8. Click continue ✔.

9. In the dashboard, set the height to **8** (Figure 7.3).

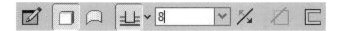

Figure 7.3

10. Complete the feature ✓.

11. View in the standard orientation using Ctrl + D.

    The completed part now looks as shown in Figure 7.4.

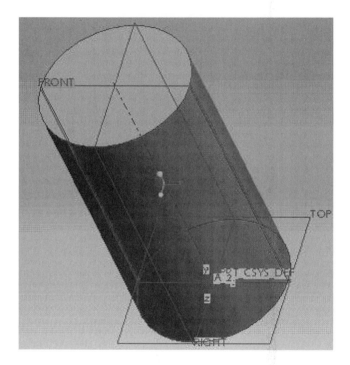

Figure 7.4

12. Save the file.

CREATING THE BOX

1. In the Pro/ENGINEER window, start a new part and name it *asm_part3*. Accept the defaults.

2. From the standard menu, click **Insert → Extrude**.

3. In the dashboard at the bottom of the screen, click the Sketcher Tool ⬚.

   (Version 2.0) Select `Placement` → `Define...`

4. In the Graphics window, click the TOP plane as the sketching plane and click Sketch to accept the defaults. Then click **Close**.

5. Using the rectangle tool ⬚, sketch a rectangle centered at the intersection of the FRONT and RIGHT planes following these steps: ⬚ → Click anywhere on the second quadrant → Drag mouse outwards to the fourth quadrant → Left mouse-click to place the rectangle creation → Middle mouse-click to end rectangle creation.

6. Middle mouse-click to enable the selection tool ⬚, or click the selection tool ⬚ from the tool bar.

7. Use the Modify Tool ⬚ to proportionately scale the sides as follows.

   A. Draw a window around the sketch (using the pointer icon).

   B. Click ⬚.

   C. Check the **Lock Scale**.

   D. In the Modify Dimensions window, click the entry corresponding to the vertical height of the rectangle and enter **10** as its value.

   E. Check the Continue button ✓.

   F. Adjust the remaining sides of the rectangle to the dimensions shown in Figure 7.5.

8. Click continue ✓.

9. In the dashboard, set the height to 1 (see Figure 7.6).

10. Complete the feature ✓.

11. View in the standard orientation using Ctrl + D.

    The completed part now looks as shown in Figure 7.7.

12. Save the file.

CREATING THE COVER

1. In the Pro/ENGINEER window, start a new part and name it *asm_part4*. Accept the defaults.

2. From the standard menu, click **Insert** → **Extrude**.

3. In the dashboard at the bottom of the screen, click the Sketcher Tool ⬚.

   (Version 2.0) Select `Placement` → `Define...`

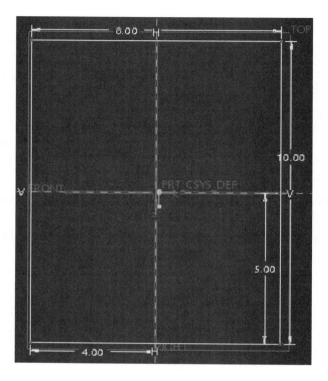

Figure 7.5

Figure 7.6

4. In the Graphics window, click the TOP plane as the sketching plane and click **Sketch** to accept the defaults. Then click **Close** in the **Reference** box.

5. Using the line and circle tools in the Sketcher, sketch a section shown in Figure 7.8.

   Note that the faint dimensions – called weak dimensions – are arbitrary and may be different on your screen.

6. Middle mouse-click to enable the selection tool ▮, or click the selection tool ▮ from the tool bar.

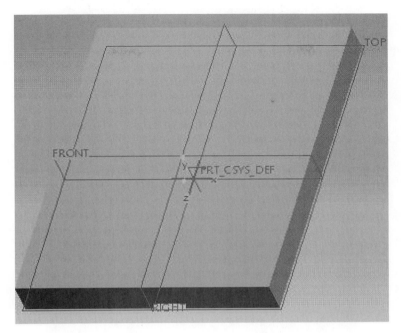

Figure 7.7

7. Use the Modify Tool ⌐ to proportionately scale the sides as follows.

   A. Draw a window around the sketch (using the pointer icon).

   B. Click ⌐.

   C. Check the **Lock Scale**.

   D. Enter **10** as the diameter of the circle.

   E. Check the Continue button ✔ from the modify tool dialog box.

   F. Adjust the remaining sides of the rectangle to the dimensions shown in Figure 7.9.

8. Click Continue ✔. In the dashboard, set the height to **4**.

9. Complete the feature ✔.

10. View in the standard orientation using Ctrl + D. The part now looks as shown in Figure 7.10.

11. Using extrusion commands as above, but with *material removal*, create a hole inside the part, of half the diameter and half the depth of the cylindrical part, so that the completed part becomes as shown in Figure 7.11.

12. Save the file.

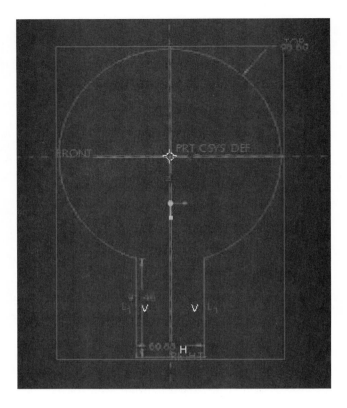

Figure 7.8

THE ASSEMBLY PROCESS
INSERTING A BUSHING INTO ONE END OF THE CYLINDER

1. Start a new file. In the **New** window, ensure that the *Assembly* option is checked as shown in Figure 7.12.

2. Name the assembly *assembly1*. Accept the default options by clicking **OK**.

3. In the tool bar to the right of the Graphics window, click Add Component .

4. In the Open window, select *asm_part1* (the bushing we created earlier).

5. In the **Component Placement** window, click **Assemble Component at default location**, as shown by the *white* cursor in Figure 7.13.

 Note: This action makes the incoming origin (the inherent origin of the bushing) coincident with the assembly origin. This first object that we have placed is also grounded, which means it cannot be moved. Subsequently placed components can move relative to it, however.

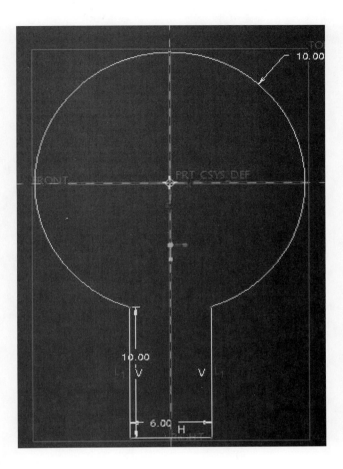

Figure 7.9

6. Click **OK**.

7. Click Add Component ⬚. Select *asm_part2* (the cylinder). Click **Open**.

8. In the **Component Placement** window, the **Placement Status** shows no constraints and the two components are placed side by side as shown in Figure 7.14.

9. Click the **Constraints** → **Type**, choose **Insert** as shown in Figure 7.15.

10. The dashboard asks for surface. Click outer surface of cylinder and then the inner surface of the bushing.

    The cylinder is inserted into the other cylinder as shown in Figure 7.16.

    Notice that the Placement Status shows **Partially Constrained**. That is because the cylinder can still be moved along its axis: Inserting the cylinder into the bushing does not preclude that. We can, however, constrain the axial

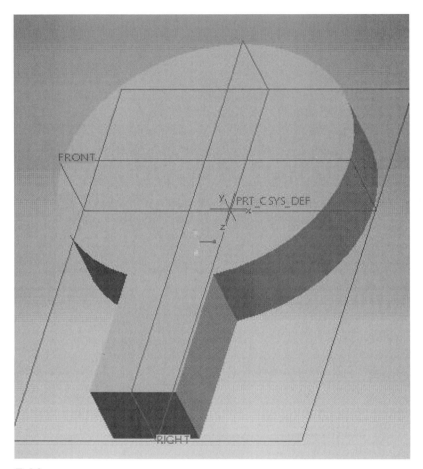

Figure 7.10

movement of the cylinder relative to the bushing by adding one more constraint – aligning the surfaces might do that.

11. Add another constraint by clicking ✦ in the **Component Placement** window.

12. Click on **Automatic** and select **Align**. Select the TOP plane of the cylinder and the TOP plane of the bushing.

    Note that the **Placement Status** indicates **Fully Constrained**, subject to assumptions. The assumptions here mean that the cylinder can still rotate relative to the bushing.

13. Click **OK**.

The fully constrained assembly now looks like that shown in Figure 7.17.

Figure 7.11

USING THE MATE CONSTRAINT

1. Click Add Component 🖳 . Select *asm_part3* (the box). Click **Open**.

2. Return to the standard view, if it is not already so, by using Ctrl + D.

3. In the **Component Placement** window, the **Placement Status** shows no constraints. The assembly and the component are placed side by side as shown in Figure 7.18.

4. Click the **Constraint** → **Type**, choose **Mate**.

5. Spin the entities by holding down the middle mouse and dragging to reveal the underside surfaces for selection. Select the underside surfaces of the component and the assembly as the mating surfaces.

Figure 7.12

The resulting package when spun around is shown in Figure 7.19.

Notice that the **Constraints** window show **Mate Coincident** and that surfaces do not actually stay in contact. However, the box is turned around so that the selected surfaces are facing each other. Coincident mating makes the surfaces coincident with a common plane, though not actually touching. That is all mating does. To actually make the surfaces touch, additional constraints are required. An example of a constraint that will make the surfaces touch is obtained if we mate the surfaces of appropriate planes in the component as well as in the assembly, in addition to our previous constraint.

ADDING ANOTHER CONSTRAINT

1. Select + to add another constraint.

2. Select the **Mate** option as shown in Figure 7.20.

3. Select the Assembly FRONT plane and select the *ASM_PART3* FRONT plane to mate these two surfaces together. Confirm correct selection from the dashboard display.

4. Enter **0** as the offset distance (if prompted).

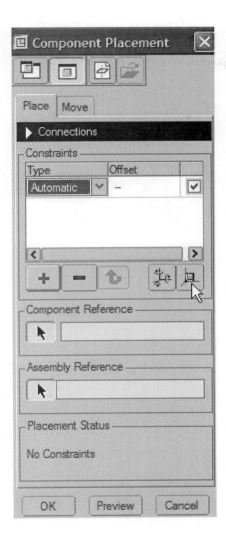

Figure 7.13

5. Repeat Step 3 above for the RIGHT planes.

6. The resulting package looks like that shown in Figure 7.21.

PACKAGING THE PLATE USING MATE AND ALIGN CONSTRAINTS

1. Click Add Component ![icon]. Select *asm_part4* (the cover). Click **Open**.

2. In the **Component Placement** window, the **Placement Status** shows no constraints. The assembly and the component are placed side by side as shown in the Figure 7.22.

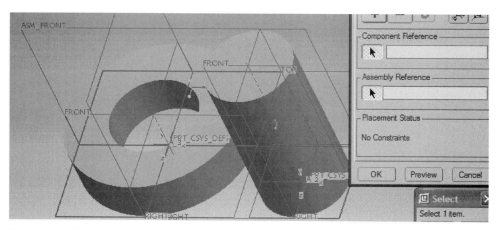

Figure 7.14

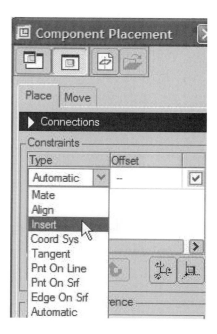

Figure 7.15

3. Click the **Constraint** → **Type**, choose **Mate** as shown in Figure 7.23.

4. Select the faces A and B of the assembly and component as shown in Figure 7.24.

5. Click the + sign to add another constraint. Select **Align** → **Constraint**.

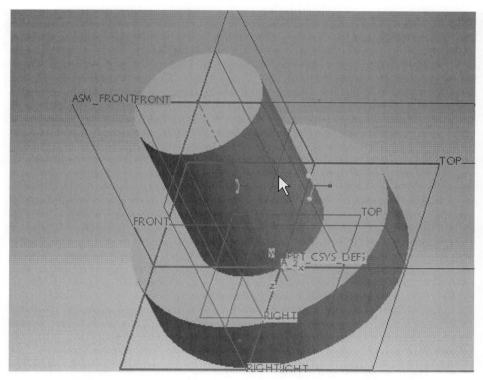

Figure 7.16

6. Select the inner surface of hole in *ASM_PART4* component and the outer surface of the *ASM_PART2* in the assembly, so that Pro/ENGINEER aligns the respective axes of these surfaces.

7. Click **OK**.

8. Toggle off the datum plane, datum axis, datum points, and coordinate system using these tools, ⬜ ⬜ ⬜ ⬜, if necessary.

   The package now looks like that shown in Figure 7.25, after an appropriate spinning of the display.

ENHANCING WITH COLORS

1. From the menu bar, select **View → Color and Appearance**.

2. In the Appearance Editor window, click the **Add new appearance** + as shown in Figure 7.26.

3. Under **Properties** tab, ensure that **Basic** is selected. Click the space to the right of **Color**, as shown in Figure 7.27.

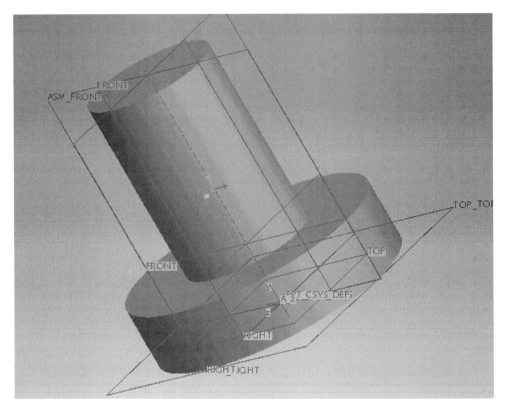

Figure 7.17

4. In the **Color Wheel** that pops up, click a suitable color – red – for example (Figure 7.28).

5. Click close.

6. Repeat steps 2–4 to add more colors to create a color palette such as shown in Figure 7.29.

7. Now let us apply the colors to the models.

   A. Click a color in the palette.

   B. Click on the drop-down arrow and select **Components** to refresh the selection and allow a choice of desired component.

   C. In the Graphics window, click a component (the bushing, for example).

   D. Click **OK** → Click **Apply**.

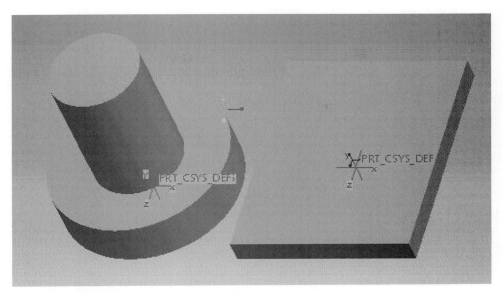

Figure 7.18

The selected component acquires the selected color.

8. Repeat until the assembly colors are as desired (Figure 7.30).

EXPLOSION OF AN ASSEMBLED PART

In order to enhance the visibility of the assembled parts, it may be necessary to separate the parts out. This is called explosion.

1. Create and assemble the three simple parts specified in Figure 7.31, Figure 7.32, and Figure 7.33. The *BASE* part is made of a cuboid 10 × 10 × 5. The hole is centered on the cuboid and is 5 units in diameter. The *HOLE* is a cylinder 10 units tall and 5 units in diameter. The *CAP* had internal diameter and external diameter of 5 and 8 units, respectively. The height of the cap is 4 units.

2. Create a new assembly file and name it *explosion_assm*. Assemble the parts so that the assembly looks like that shown in Figure 7.34.

3. To explode the view, using the standard menu bar, click **View → Explode → Explode View**.

4. Again, click **View → Explode → Edit Position**.

5. Under motion type, verify that **Translate** is selected. Under **Motion Reference**, select **Plane Normal**. Select the top surfacez of the *BASE* so that components can be moved normal to the selected surface.

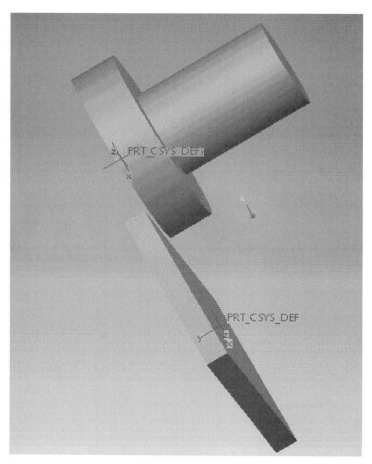

Figure 7.19

6. Click on component and drag. Motion will be normal to the plane, as selected in Step 5 above. Arrange the components so that the exploded arrangement is similar to that shown in Figure 7.35.

7. Click **OK** to accept the arrangement.

We must, however, preserve the view so that this arrangement can easily be recalled.

8. Click the **View Manager** icon 🔲 on the standard menu bar, typically to the left of the no-shade icons. The **View Manager** icon is displayed.

9. Click New and type *Myexplode* as the name of the view → Hit the Enter key → **Close**. The exploded view is now saved as *Myexplode*. This exploded view will be used in the drawing mode.

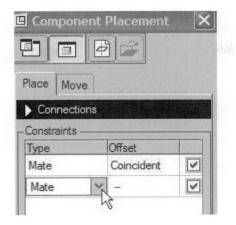

Figure 7.20

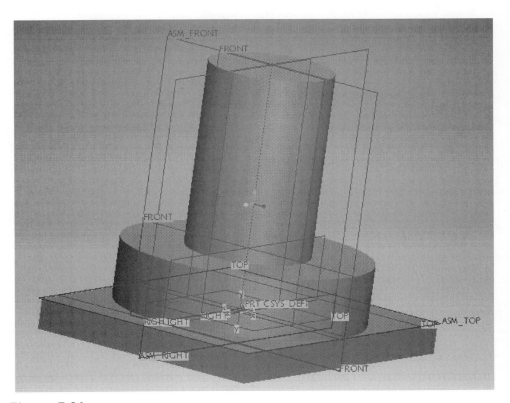

Figure 7.21

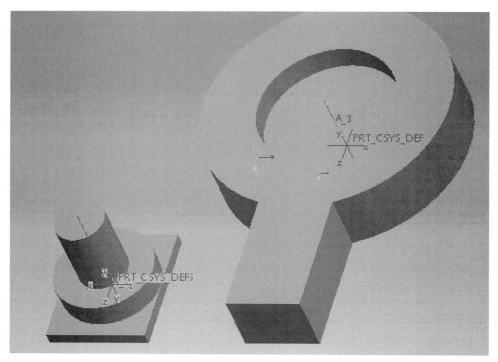

Figure 7.22

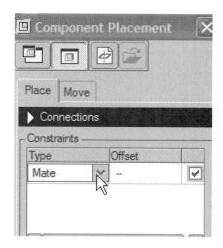

Figure 7.23

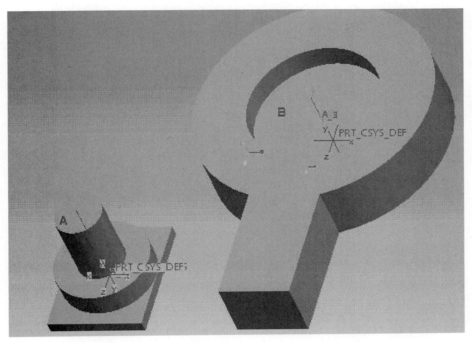

Figure 7.24

10. Create a new Pro/ENGINEER drawing and name it *explosiondrg*. Make the assembled part *explosion_assm* the default model. Use the **c-drawing** template.

11. Click Insert drawing view ▣ → **General** → **Full View** > **Exploded** > **No Scale** > **Done**.

12. The dashboard requests for the location of the center point. Click to the north-east of the default view. Ensure that Myexplode is selected in the **Menu Manager** that appears → **Done** → **OK**. The exploded view is placed in the drawing file, as shown in Figure 7.36.

BILL OF MATERIALS (BOM)

The Bill of Materials summarizes the components of an assembly. It is a helpful presentation tool. The creation of BOM starts with the creation of a table.

1. With the *explosiondrg* file opened, click **Table** → **Insert** → **Table** → **Descending** → **Rightward** → **By Num Chars** → **Pick Pnt**.

2. Read the dashboard and note that the upper left of the table is required. Click at a suitable location to indicate the upper left of the table.

3. Click to the right to specify the width of the first column and again to the right to specify the width of the second column → Middle mouse-click to end column width specification.

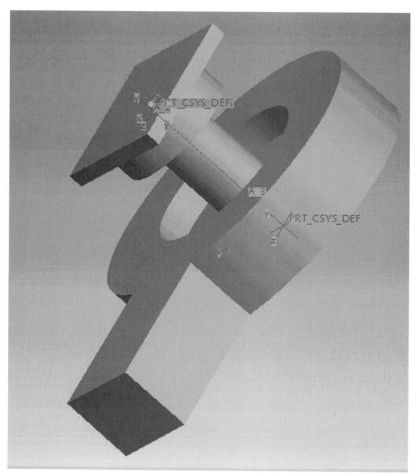

Figure 7.25

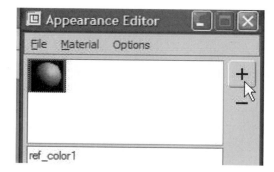

Figure 7.26

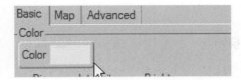

Figure 7.27

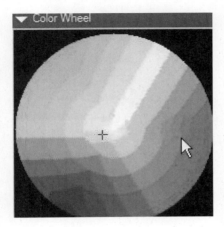

Figure 7.28

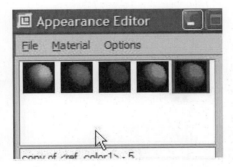

Figure 7.29

4. Click to the bottom of the upper left corner to specify the height of the first row and again to the bottom to specify the column width of the second row → Middle mouse-click to end the row height specification. The two rows hereby specified are sufficient to start with. The first row is static and takes the headers. The second row is dynamic: Pro/ENGINEER populates the cells with the parameters specified.

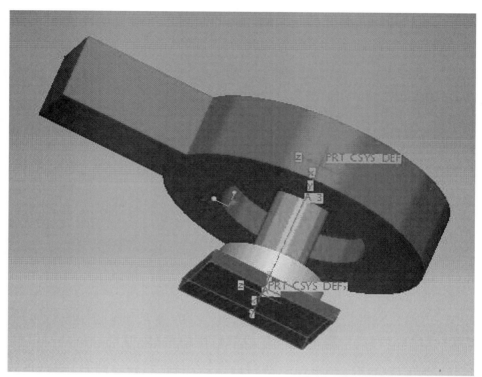

Figure 7.30

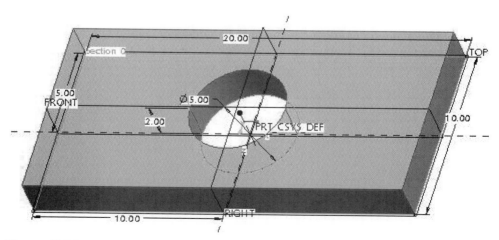

Figure 7.31

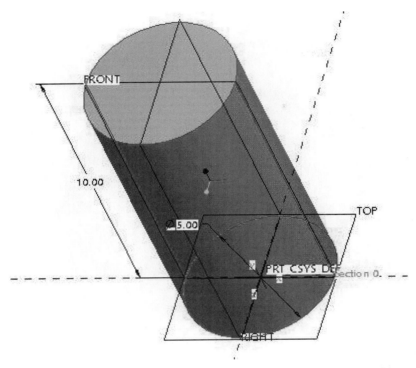

Figure 7.32

5. Double-click the upper left cell. The **Note Properties** window appears → Type *Index* in text area → **OK**.

6. Double-click the upper right cell. The **Note Properties** window appears → Type *Part Name* → **OK**. The table should now look like that in Figure 7.37.

7. Click **Table** → **Repeat Region** → **Add**. The dashboard requests for the corners of the region.

8. Click the lower left cell → Click the lower right cell. Both cells should now be highlighted. Ensure that the dashboard displays the message "Repeat region successfully created". If not, it may be better to delete the table so far created and repeat steps 1–8.

9. Click Done. We are now ready to add the BOM parameters which Pro/ENGINEER uses to automatically populate the table cells.

10. Double-click the left cell of the repeat region below the row where the typed texts have been placed. The Report Symbol window opens. The window gives options about which Pro/ENGINEER can extract information.

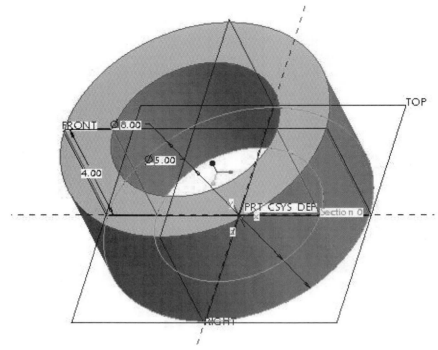

Figure 7.33

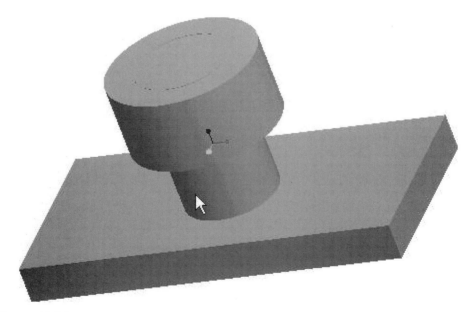

Figure 7.34

Figure 7.35

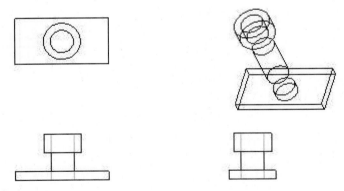

Figure 7.36

Index	Part Name

Figure 7.37

11. Click **rpt (report)** → **index**. Click the lower right cell → **asm** → **mbr** → **name**. This will fetch the assembly member name.

12. Select **Table** → **Repeat Region** → **Update Tables**. The table is dynamically populated to show the information associated with the specified parameters in Step 12.

BOM BALLOONS

1. Click **Table** → **BOM Balloons** → **Set Region**. Click on the region, beneath the row where title row.

2. Click **Create Ballon** → Select the general view.

3. The Balloons are added to the view, as in Figure 7.38.

4. Select any arrow of the balloons and arrange as desired.

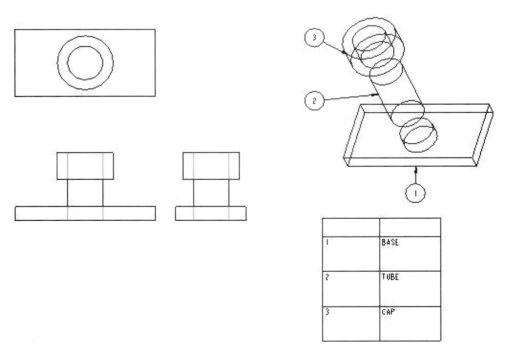

Figure 7.38

EXERCISE

1. Use the components created earlier to create the assembly shown in Figure Q7.1.

2. Create the assembly in Figure Q7.2a. The hammer head to shaft is fully constrained. The base to the shaft may be partially constrained but symmetrically placed.

   The cross section of the hammer looks as shown below in Figure Q7.2b.

   Use suitable dimensions for the length of the shaft and the radii of the bushing.

   Generate a drawing of the hammer. It should look like that shown in Figure Q7.2c.

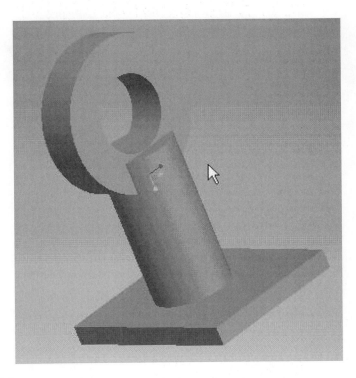

Figure Q7.1

Figure Q7.2a

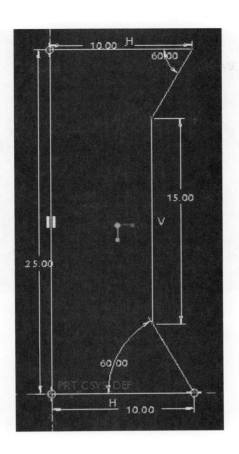

Figure Q7.2b

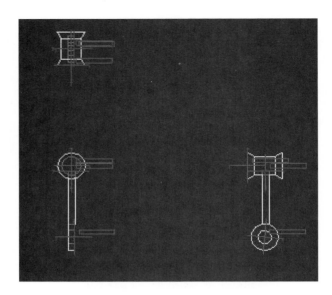

Figure Q7.2c

CHAPTER 8

Feature Operations

THE BASICS

Feature operations apply to the created features of a part. That means the process of extrusion, revolution and similar processes have been completed and the designer wishes to use an existing feature to create others.

The goals of this chapter include creating patterns, copying, and mirroring.

CREATING THE BASE BOX

Create the box with sides **20 × 10** and height **2** units as follows:

1. Start a new part file named *operations1*. Accept the defaults.

2. Using the features tool bar, click ⬚ → ⬚.

3. In the **Sketch** window, click TOP in the Graphics window as the sketching plane and accept the defaults.

4. Create the sketch shown in Figure 8.1.

5. Continue the sketching by clicking ✔. Set the height to **2.00** units, by setting the dashboard as shown in Figure 8.2.

   The completed part looks like that shown in Figure 8.3.

CREATING THE TOP BOX

We are going to create a smaller box on top of the box.

1. Click ⬚ → ⬚ → Click the top surface of the created box in the Graphics window as the sketching plane → **Sketch** → **Close**.

2. Change to the wireframe display by clicking ⬚ for easy viewing of the parts.

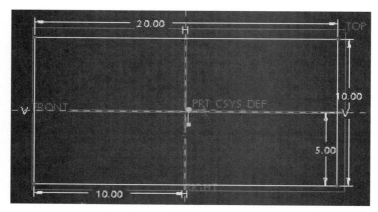

Figure 8.1

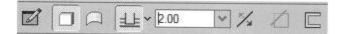

Figure 8.2

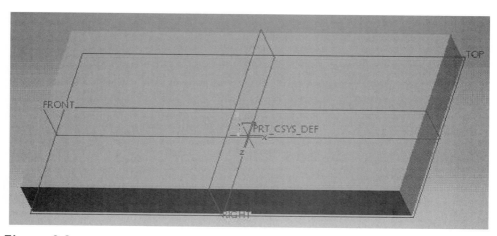

Figure 8.3

3. Draw a rectangle at the right corner to create a sketch similar to that shown in Figure 8.4.

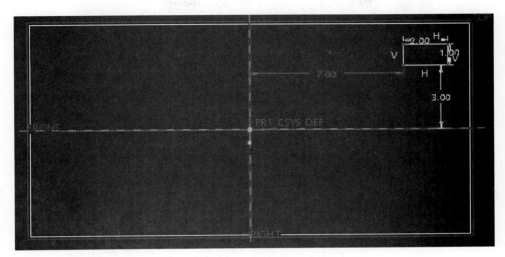

Figure 8.4

4. Click Continue ✔.

5. Set the height as **1.00** in the dashboard as shown in Figure 8.5.

Figure 8.5

6. Click apply ✔. View in the standard orientation (Ctrl + D) and turn on the Shading mode ◻. The part now appears as shown in Figure 8.6.

RECTANGULAR PATTERNING

We wish to pattern the smaller box on top of the bigger one.

1. Select the small box in the Graphics window so that it turns red, if it is not already so.

2. Using the features tool bar, click the **Pattern Tool** ▦. Click **Dimensions** tab in the dashboard, as indicated by the cursor position in Figure 8.7.

The Dimension pop-up window shows that we have two options of the directions of dimensioning: **Direction 1** and **Direction 2**. The window also

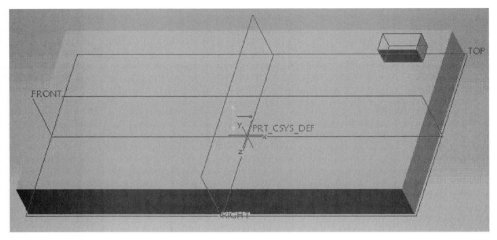

Figure 8.6

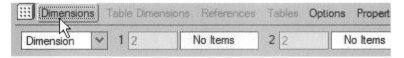

Figure 8.7

shows that we can vary the dimension of the patterning by relations. We need to, however, tell Pro/ENGINEER which is **Direction 1** and which is **Direction 2**.

3. Click in the **Direction 1** window. Then click the dimension **7** in the Graphics window to associate this with **Direction 1**. The existing length **7** is highlighted. Type **−4** to set the increment to **4** in the Direction 1 dialog. You can also set the increment to **−4** in the Dimensions tab window.

4. In the dashboard, set the number of items to **5**, as indicated by the cursor position in Figure 8.8. This includes the original in the 5.

Figure 8.8

5. Complete the feature ✔.

The completed part now looks like that shown in Figure 8.9.

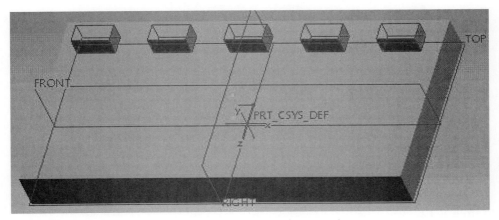

Figure 8.9

6. In the Model Tree, right-click on the pattern icon → **Edit Definition**, as indicated by the cursor position in Figure 8.10.

Figure 8.10

7. In the dashboard, click the **Dimensions** tab [Dimensions].

8. Click the window beneath the **Direction 2** label, as shown in Figure 8.11.

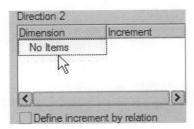

Figure 8.11

9. Click the **3** dimension in the Graphics window and enter the increment −**2**, as indicated by the cursor position in Figure 8.12.

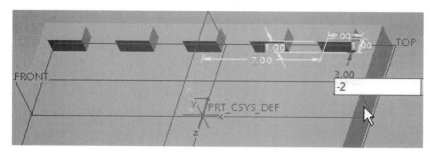

Figure 8.12

10. Enter **4** as the number of items in direction 2 as shown in Figure 8.13.

Figure 8.13

11. Complete the feature ✔. The completed part now becomes like that shown in Figure 8.14.

12. Save the file.

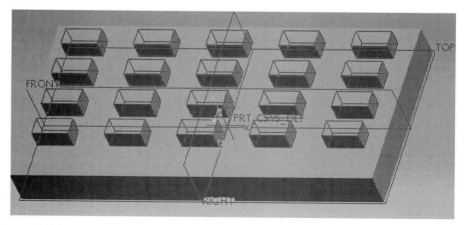

Figure 8.14

RADIAL PATTERNING

Note that Radial patterning requires the existence of an axis of rotation and an *angular dimension* to vary. Axis of rotation may be created by using the axis tool and selecting two planes which intersect to form the axis. Angular dimension may be created by using feature operations followed by rotation. Pro/ENGINEER stores the angle of rotation that may be used to specify radial patterning. The steps are explained below.

TASK I

It is convenient to re-use our existing file.

1. Open the file *operations1*, if it is not already opened.

2. **File** → **Save a Copy**. Name the part *operations2* → **OK**, as shown in Figure 8.15. Close *operations1* file so as to open the newly named *operations2* file.

Model Name	OPERATIONS1.PRT
New Name	operations2
Type	Part (*.prt)

Figure 8.15

3. Open *operations2* file.

4. Right-click on patterns → **Delete Pattern**.

Caution: When you right-click on Patterns, there is *Delete* and there is *Delete Pattern*. Delete Pattern will keep the original feature, but delete only the pattern.

The part reduces to that shown in Figure 8.16.

5. Edit the placement of the smaller box to bring it closer to the origin at the intersection of the FRONT and the RIGHT planes, so that the sketch becomes as shown in Figure 8.17 (Right-click on the smaller box in the Model Tree → **Edit Definition** →). (Version 2.0) Select Placement → **Edit** → **Sketch** → **OK**. Using the standard menu bar, click the Wireframe mode for clarity. Resize as necessary. (Location is approximate.)

6. Click continue → **OK** → .

7. View in the standard orientation, so that the part becomes like that shown in Figure 8.18.

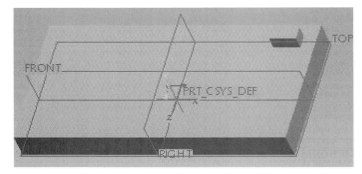

Figure 8.16

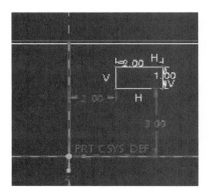

Figure 8.17

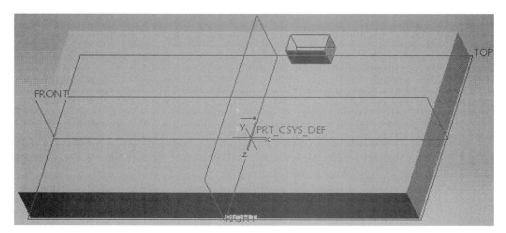

Figure 8.18

We need to create an angular dimension. This can be accomplished by *copying* and *rotating* the existing feature – the small box. First we need to create an axis of rotation.

8. Create an axis representing the intersection of the FRONT and RIGHT planes.

   A. Click the **Datum Axis Tool** .

   B. Holding down the control key, select the FRONT and RIGHT planes. Observe that the FRONT and the RIGHT planes are listed as **References** in the **Datum Axis** window.

   C. Click **OK**.

   D. Notice that the Model Tree shows the creation of an axis named **A_1**.

9. Using the standard menu bar, click **Edit** → **Feature Operations** → **Copy** → **Move** → **Select** → **Independent** → **Done**.

10. Select the small box in the graphics window → **OK** → **Done** → **Rotate** → **Crv/Edge/Axis**.

11. Select the **A_1 axis** → **OK**.

12. Enter **30** as the angle of rotation → Accept the value → **Done Move**.

   The group variable dimensions that then pops up allows us to change the dimensions of our copied object. By checking the appropriate variable, we can change one or more dimensions of our copy. We do not want to change any dimensions at this time.

13. Click **Done** → **OK** → **Done**.

   The copied feature and the original now look like that shown in Figure 8.19.

   We can now pattern the copied member. We could delete the original small box, if we want. This is possible because during the copy operations, we had selected the independent option, so that the copied box and the original are independent. Occasionally, a pattern operation will fail and Pro/ENGINEER may have to delete the pattern along with the original! Therefore, we want to complete our pattern operation successfully before we delete the original.

14. With the copied box selected, click the **Pattern Tool** .

15. Select the angular dimension **30** degrees and enter an increment value of **60** degrees, as shown in Figure 8.20. *Make sure to hit the Enter key to accept the 60 value.*

16. In the dashboard, enter **6** as the number of items for **Dimension 1**, as shown in Figure 8.21.

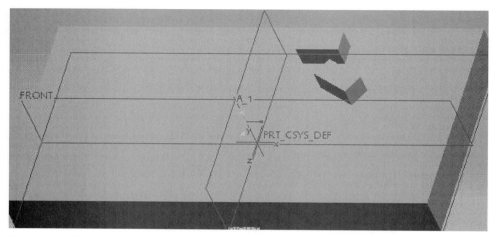

Figure 8.19

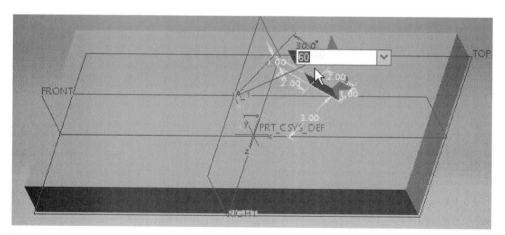

Figure 8.20

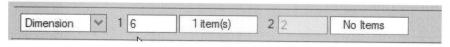

Figure 8.21

17. Complete the feature ✓.

18. Our pattern works. Select the original box and delete it (see Figure 8.22).

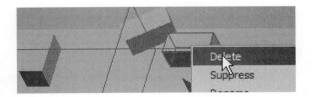

Figure 8.22

Our part now looks like that shown in Figure 8.23.

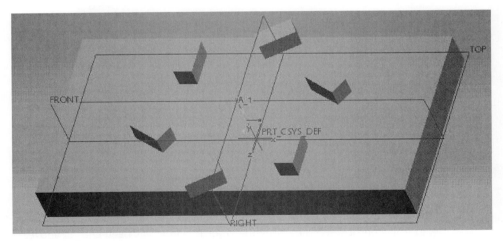

Figure 8.23

19. Save the file.

FURTHER EXAMPLE OF RADIAL PATTERNING, USING A HOLE

TASK 2

1. Open the file *operations2*, if it is not already opened.

2. **File → Save a Copy**. Name the part *operations3* → **OK**, as shown in Figure 8.24.

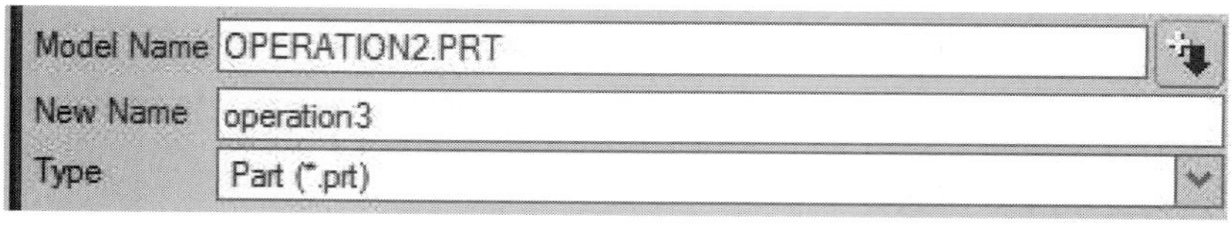

Model Name	OPERATION2.PRT
New Name	operation3
Type	Part (*.prt)

Figure 8.24

3. Close *operations2* and open *operations3*.

4. In the Model Tree window, the copied features are listed as a group. Right-click on the group and ungroup it.

5. Right-click on patterns → **Delete**. (This will delete the pattern and the original.)

6. Using the menu bar, **Insert** → **Hole**. Click **Placement** in the dashboard. In the **Placement** box, click the **Primary** box. In the Graphics window, click the top surface of the box feature. This provides the surface on which the hole will be drilled.

7. Again, in the **Placement** box, change the mode to radial as shown in Figure 8.25.

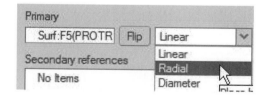

Figure 8.25

Radial hole placement is required to provide the angular dimension to vary for patterning purposes. That is why the "radial" option has been chosen above.

8. Click the **Secondary references** box and then select the axis A_1 in the Graphics window or in the Model Tree window.

9. Holding down the Ctrl key, select the RIGHT plane, so that the placement is now fully constrained.

10. Set the dimensions as shown in Figure 8.26.

11. Complete the feature ✓. The completed part looks like that shown in Figure 8.27.

   We now want to pattern the hole around the A_1 axis.

12. Select the hole feature and click the **Pattern Tool** ▦.

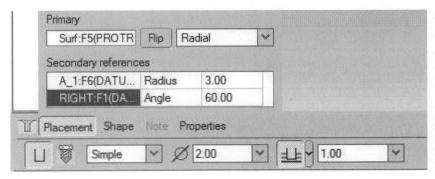

Figure 8.26

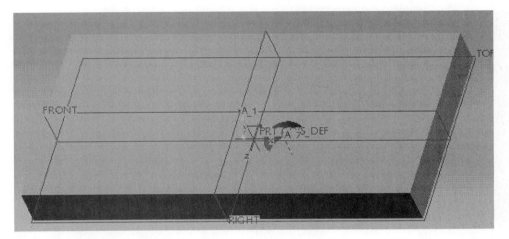

Figure 8.27

13. Click **Dimensions** → Click the white space below **Direction 1**. In the graphics plane, click the angular dimension of **60** degrees and enter **45** degrees as shown in Figure 8.28.

The increment can also be set in the **Direction 1** drop-down menu as shown in Figure 8.28.

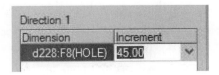

Figure 8.28

The steps above can also be realized by clicking the angular dimension of 60 degrees in the Graphics window and setting it to 45 directly. However, clicking Dimensions first enables more options.

14. In the dashboard, set the number of items to 8, as shown in Figure 8.29.

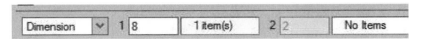

Figure 8.29

15. Complete the feature ✔.

The completed part now looks like that shown in Figure 8.30.

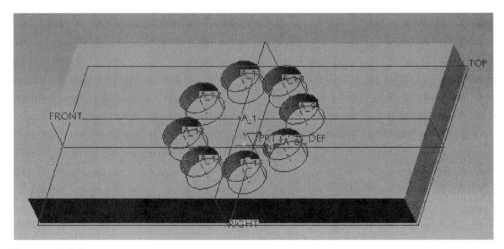

Figure 8.30

MODIFYING THE BASE HOLE OF THE PATTERN

The base hole of the pattern is the original hole that was patterned.

1. Double-click on the **+** sign behind Patterns in the Model tree ⊞ ▣ Pattern (COPIE. We see a series of hole Ids. The first one is the base hole.

2. Right-click on the base hole (the first Hole Id) and select **Edit**, as indicated by the cursor position in Figure 8.31.

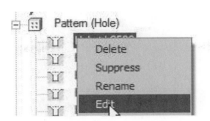

Figure 8.31

3. By clicking on the respective values in the Graphics window, change the radial distance of the base hole from **3** to **2** and the radius of the base hole diameter from **2** to **1**, as indicated by the cursor position in Figure 8.32.

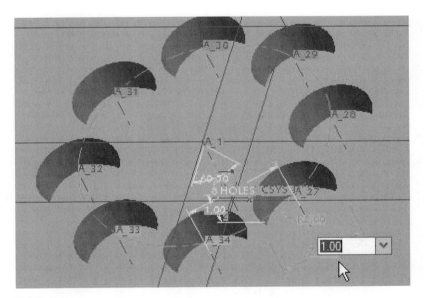

Figure 8.32

4. Regenerate the model 🔳. The pattern now appears shrunk as shown in Figure 8.33.

5. Minimize the Pattern by clicking the – behind pattern in the Model Tree, if necessary.

6. Right-click on Pattern in the Model Tree window and select Edit Definition, as indicated by the cursor position in Figure 8.34.

7. Click **Dimensions** in the dashboard → Click the white space below **Direction 2**.

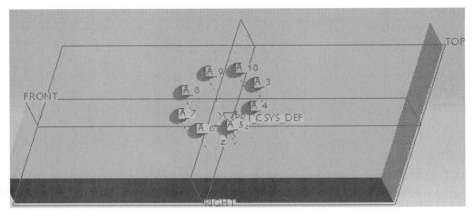

Figure 8.33

Figure 8.34

8. Select the radial dimension of the hole as the dimension along **Direction 2**. Accept the default increment of **2** units as shown in Figure 8.35.

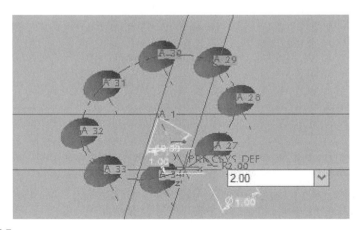

Figure 8.35

9. In the dashboard, keep the number of items along **Dimension 2** to be **2**, as shown in Figure 8.36.

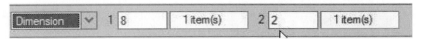

Figure 8.36

10. Complete the feature ✓. The completed figure now appears as shown in Figure 8.37. Notice the variations along the two directions: angular variation and radial variation.

11. Save the file.

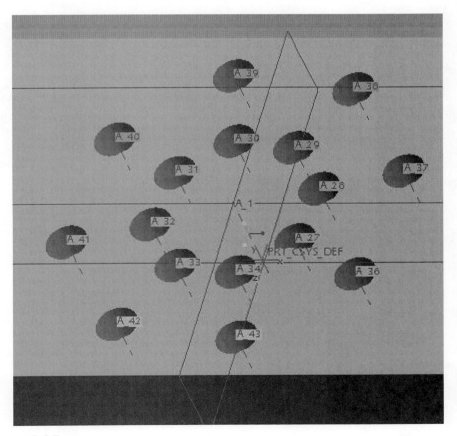

Figure 8.37

MIRRORING

1. Create the following part using the steps shown.

    A. Open *operations1*.

    B. **File → Save a Copy**. Name the part *operations4* → **OK**.

    C. Open *operations4*.

    D. Right-click on patterns → **Delete Pattern**.

    The resulting part reduces to that shown in Figure 8.38.

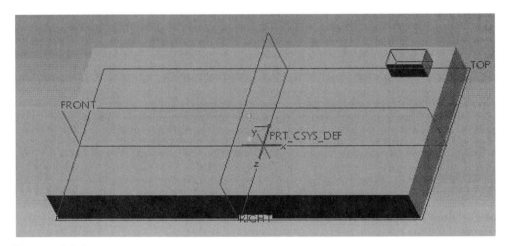

Figure 8.38

2. Using the standard menu, click **Edit → Feature Operations → Copy → Mirror → Select → Independent → Done**.

3. Select the small box, so that it turns red → **OK → Done**.

4. Click the RIGHT plane as the plane of reflection → **Done**.

    The completed part looks like that shown in Figure 8.39.

5. Again, using the standard menu, click **Edit → Feature Operations → Copy → Move → Select → Independent → Done**.

6. Select the copied small box, so that it turns red → **OK → Done → Translate**.

7. Click the FRONT plane → **Okay** → Enter **4.0** → ✓ → **Done Move → Done → OK → Done**.

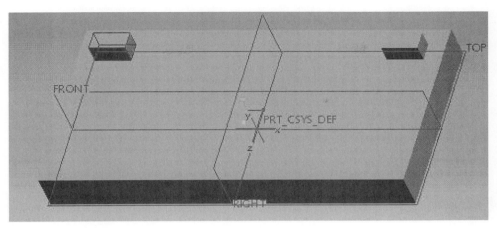

Figure 8.39

The completed part now looks as shown in Figure 8.40.

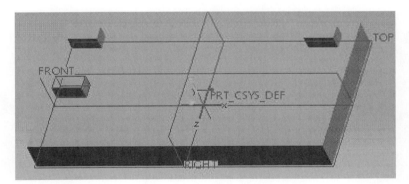

Figure 8.40

8. Repeat steps 5 and 6 above, selecting the original small box at the right corner.

9. Click the FRONT plane → **OK** → Enter **8.0** → ☑ → **Done Move**.

10. Check the dimensions to change as shown in Figure 8.41.

    Notice that Pro/ENGINEER highlights the corresponding sides to the checked dimensions.

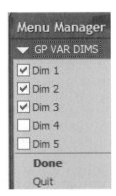

Figure 8.41

11. Click **Done**. Enter **2,2,2** as the new dimensions of the checked sides → **OK**. The completed part now looks like that shown in Figure 8.42.

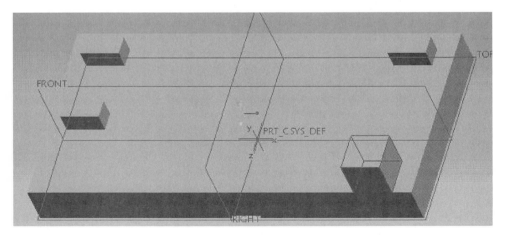

Figure 8.42

EXERCISE

1. Experiment with other feature operations.

2. Create the 3D model of the object shown in Figure Q8.2. Four brackets are located within the hollow box. Your drawing should include the creation of a bracket and then patterning the bracket to create the other brackets. Use free but proportionate dimensions.

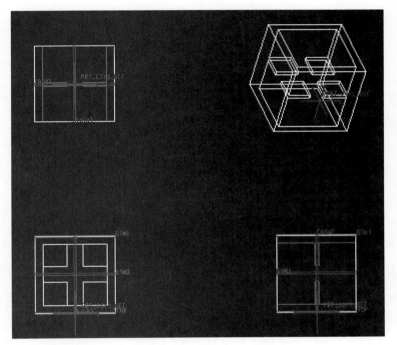

Figure Q8.2

CHAPTER 9

Parent–Child Relationship

THE BASICS

The understanding of parent–child relationship helps us carry out part modification that Pro/ENGINEER can handle. A child entity is dependent on the parent entity. A feature (child) is created relative to another surface or line (parent). Therefore, deleting the parent feature destroys the information that Pro/ENGINEER needs to uniquely place the child feature.

Relationship can exist because a feature (child) is created using the information of another feature (parent). This can occur, for example, if the child was created using a plane of the parent or using measurements relative to a parent. These are known as references. To remove relationship, therefore, one must remove these relationships.

CREATING THE BASE BOX

1. Begin a new part and name the part *parent–child*.

2. Select the extrude tool and create the sketch shown in Figure 9.1a on the TOP plane.

3. Accept the sketch by clicking ✔.

4. Use the dashboard setting for the extrusion to be similar to that shown in Figure 9.1b.

5. Complete the feature so that the created feature now looks like that shown in Figure 9.2.

6. In the Model Tree, rename the extrusion *BASE*, as indicated by the cursor position in Figure 9.3.

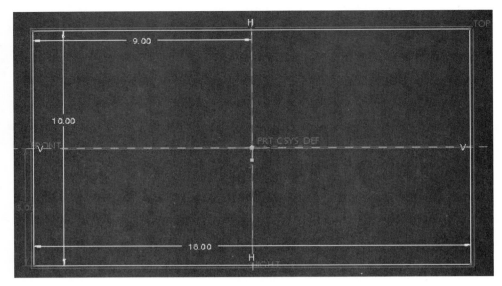

Figure 9.1a

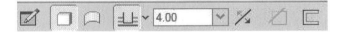

Figure 9.1b

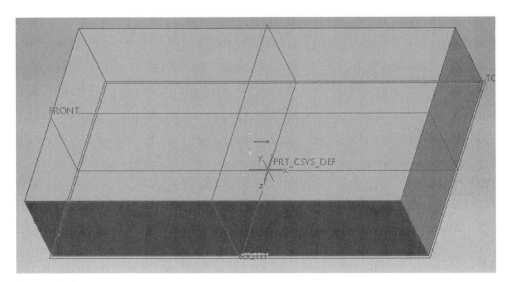

Figure 9.2

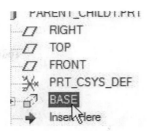

Figure 9.3

CREATING THE LINK EXTRUSION

1. Similarly, create a second extrusion with the top of the BASE extrusion as the sketching plane. The dimensions of the second extrusion are shown in Figure 9.4.

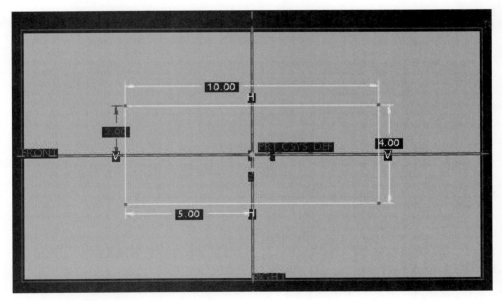

Figure 9.4

2. Use the same entries of the dashboard as the settings for the BASE extrusion.

3. Complete the feature so that the part now becomes like that shown in Figure 9.5.

4. In the Model Tree, rename the extrusion *LINK*, as shown in Figure 9.6.

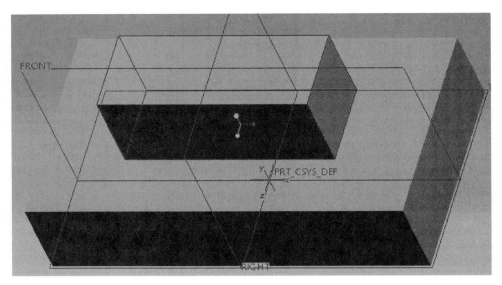

Figure 9.5

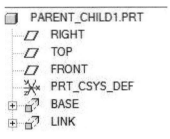

Figure 9.6

PUTTING A RECTANGULAR HOLE IN THE BASE EXTRUSION

1. Click the extrusion tool ⬚ and then go to the Sketcher. Select top of the base of the BASE extrusion as the sketching plane and accept the default.

2. Create a sketch like the one shown in Figure 9.7. Be sure to make the sketch symmetric about the horizontal line (FRONT plane) by drawing a centerline along the FRONT plane edge (or the horizontal line) and using the symmetric constraints on the top and bottom ends of the sketch's vertical line.

3. Click the Continue button ✔.

   Ensure that the removal direction is checked in the dashboard.

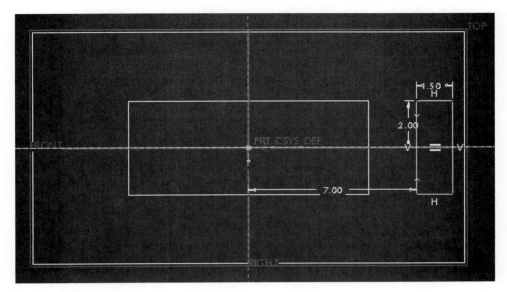

Figure 9.7

4. Complete the feature ✔. The modified part now looks like that shown in Figure 9.8.

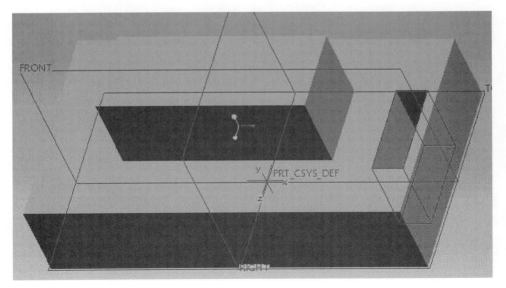

Figure 9.8

5. Using the Model Tree, rename the cut as *SLOT*, as shown in Figure 9.9.

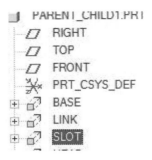

Figure 9.9

CREATING THE HEAD EXTRUSION

1. Create another extrusion, by selecting the Extrude Tool and going into the Sketcher Tool. Select the top of the *LINK* extrusion as the sketching plane. The dimensions of the second extrusion are shown in the Sketcher as shown in Figure 9.10.

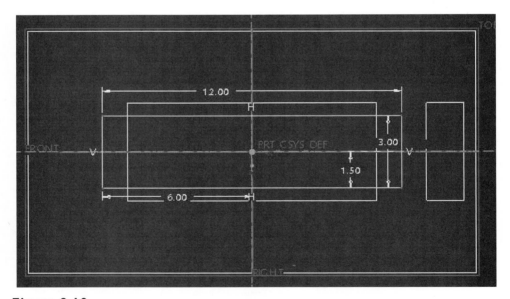

Figure 9.10

2. Use the same entries of the dashboard as the settings for the BASE extrusion.

3. Complete the feature so that the part now becomes like that shown in Figure 9.11.

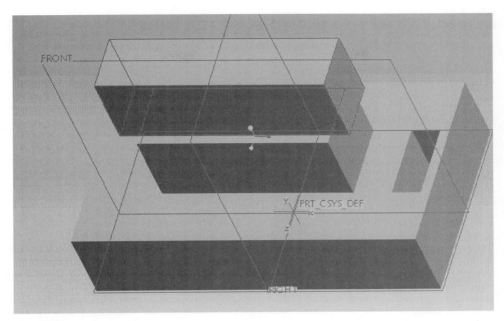

Figure 9.11

4. In the Model Tree, rename the extrusion *HEAD*, as shown in Figure 9.12.

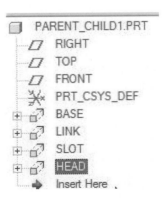

Figure 9.12

Now comes specification from the engineering team that the HEAD extrusion should be aligned with the slot in the BASE extrusion.

ALIGNING THE HEAD AND SLOT FEATURES

1. In the Model Tree, right-click on the HEAD and select **Edit Definition**.

2. Click the Sketcher.

   (Version 2.0) Select Placement → [Edit...] → **Sketch**

   The following sketcher entity (Figure 9.13) is displayed.

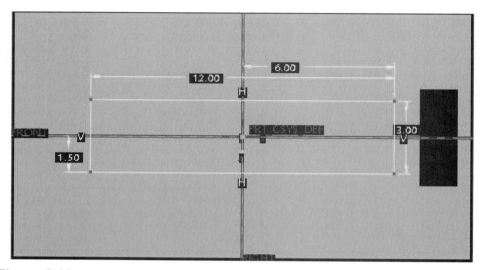

Figure 9.13

We wish to align the right vertical line of the dimensioned rectangle with the left vertical line of the slot shown as the smaller box. Although the slot is shown, we cannot click on any point on it unless we first use the **Create an entity from an edge** Tool ⬚▸.

3. Aligning the sketched rectangle and the slot will alter the **6.00** dimension presently displayed. So, let us get that out of the picture. Select the **6.00** dimension so that it turns red. Right-click on it and select delete.

4. Select ▦ tool from the Sketcher tool bar. In the **Constraints** window, click ⊙ option → Click the top right end of the dimensioned rectangle and the left side of the slot → **Close**.

   Observe that the edges of the rectangle and the slot have been aligned.

5. Click ✔ → **OK** → ✔ → Ctrl + D, to view in the standard orientation. The part now looks like that shown in Figure 9.15.

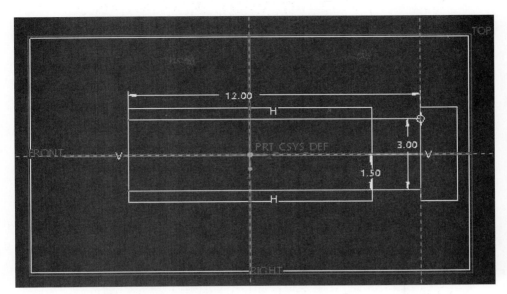

Figure 9.14

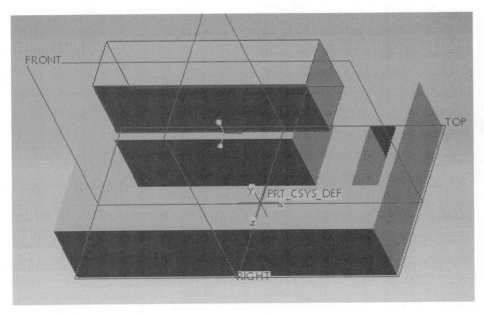

Figure 9.15

RELATIONSHIPS

1. In the Model Tree, right-click on the HEAD extrusion → **Info** → **Parent/ Child**. The **Reference Information Window** (Figure 9.16) is displayed.

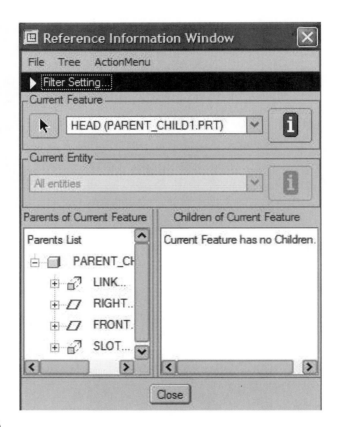

Figure 9.16

Notice that the HEAD extrusion has no children but has several parents. One of the parents is the SLOT feature. SLOT became a parent of HEAD because HEAD was aligned with SLOT. That is, HEAD is dependent on SLOT.

2. Close the Parent–Child information window.

3. In the Model Tree, right-click on the LINK extrusion → **Info** → **Parent/Child**. The following **Reference Information Window** (Figure 9.17) is displayed.

Notice that the LINK extrusion has entries in both the **Parents** List and **Children** List. The display shows that BASE is a parent of HEAD while HEAD is a child of LINK.

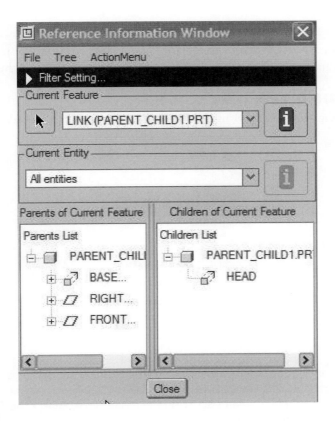

Figure 9.17

With the feature now created and also with the knowledge of parent–child relationships, let us carry out the SUPPRESS and HIDE operations. These operations are closely tied to the concept of parent–child relationships. In a nutshell, the differences are summarized below.

SUPPRESSION

To hasten regeneration, not all features need to be viewed. Suppression allows one or more features to be *removed* from the regeneration list and, thus, increases the speed of regeneration. Notice that the feature being suppressed still exists in the Model Tree, but Pro/ENGINEER simply skips its regeneration. Suppressing a parent feature also suppresses the child feature. A suppressed feature can be unsuppressed by using the Resume command. Resuming a child also resumes its parent. We note that a parent can exist without its children but a child cannot exist without its parents.

HIDING

Applicable to non-solid entities like planes and surfaces. Hiding is not affected by Parent/Child inheritance. That is, hiding a child has no effect on the parents.

SUPPRESSING THE SLOT

1. In the Model Tree, right-click on SLOT → **Suppress** (Figure 9.18) → **OK**.

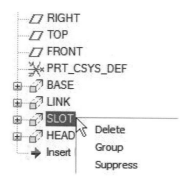

Figure 9.18

2. Observe in the Graphics window that the HEAD is gone. That is because HEAD is a child of SLOT. HEAD became a child of SLOT when its side was made coincident with a side of SLOT. In the Model Tree, HEAD and SLOT may also be gone. If they are gone, you can still display them by using the Model Tree Settings.

3. We can set the display so that even suppressed items are displayed – that way a designer can see what has been suppressed. In the Model Tree, click **Settings** → **Tree Filters**. In the **Model Tree Items** window, check **Suppressed Objects** → **Apply** → **OK**.

4. Observe now that HEAD and SLOT are displayed in the Model Tree, but preceded by a square bullet. The bullet gives indication that the items are suppressed.

5. Now undo the suppress action as follows. Using the standard menu bar, select **Edit** → **Resume** → **Last**.

HIDING A PLANE

1. Select the RIGHT plane in the Graphics windows, so that it turns red.

2. Right-click on it and select **Hide**.

3. Observe in the Model Tree that RIGHT plane is grayed out.

4. To view the hidden entities, right-click on the object in the Model Tree → **Unhide**.

EDITING RELATIONSHIPS

Let us remove the parent–child relationship between SLOT and HEAD feature.

1. In the Model Tree, right-click on the HEAD → **Edit Definition** → ▣.

   (In Version 2.0) Select `Placement` → `Edit...` → **Sketch**.

 Note: The right-hand side of the rectangle is constrained with the slot through Point on Entity constraint. What we are going to do is to create a conflicting constraint. That offers us the opportunity to remove the unwanted constraint.

2. Using the Sketcher tool bar, click ⊢⊣ and dimension the lower right corner of the HEAD's rectangle from the right-hand side of BASE protrusion (see Figure 9.19).

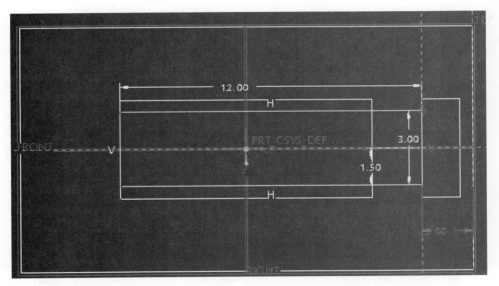

Figure 9.19

3. Pro/ENGINEER warns of a conflict and displays the window shown in Figure 9.20. Delete the Point on Entity constraint.

   We have now succeeded in constraining the right-hand side of the rectangle to the slot. We, however, need to constraint the right-hand side of the rectangle to be vertical, if it is not already so. The side of the rectangle may not remain vertical if the designer had used a series of lines, instead of using the rectangle tool, in creating the HEAD's section. In that case, click the Constraints Tool ⌖ → ⌡ → Click the right-hand side of the rectangle. This makes the right-hand side vertical.

Figure 9.20

 Note: If the right-hand side of the rectangle was already constrained vertical before editing the parent–child relationship, the steps required will be a slight variation of this. For example, the vertical constraint could be listed in the list of conflicts in the Resolve Sketch window. Since it is desired to keep the right-hand side vertical, we would keep the vertical constraint and still delete the Point on Entity constraint.

4. Edit the dimensions so that the sketch now looks like that shown in Figure 9.21.

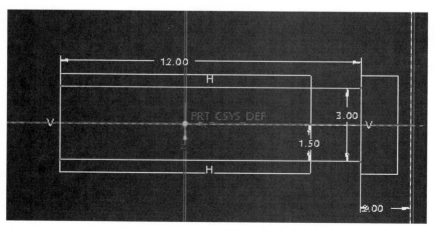

Figure 9.21

5. Click 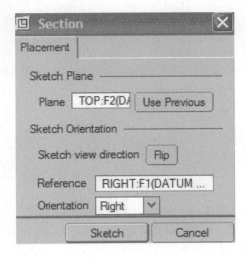 → **OK** → ✔ → Ctrl + D.

Observe that HEAD is no longer a child of SLOT.

FURTHER EXAMPLE ON PARENT–CHILD RELATIONSHIP
CREATING THE BOX FEATURE

1. Create a new part and name it *BOX*.

2. Using the features tool bar, select ⬚. Go into the Sketcher environment. In the **Placement** window that appears, click the TOP plane as the **Sketch Plane**. The rest of the Section window default to the values shown in Figure 9.22.

Figure 9.22

3. Click **Sketch** → **Close**. Then create the section shown in Figure 9.23.

4. Click continue and in the dashboard window that appears, set the thickness to **4** units, and ensure that the rest of the settings are like that shown in Figure 9.24.

5. Click continue to complete the feature creation. Using Ctrl + D to view in the standard orientation, the created feature is like that shown in Figure 9.25.

6. In the Model Tree window, right-click on the protrusion and rename it as *BOX*.

CREATING THE CUT ON THE SIDE ABCD

1. As before, using the features tool bar, click ⬚ → ⬚. In the **Placement** window that appears, click the ABCD plane. Accept the default settings of the Section window and click **Sketch** → **Close**.

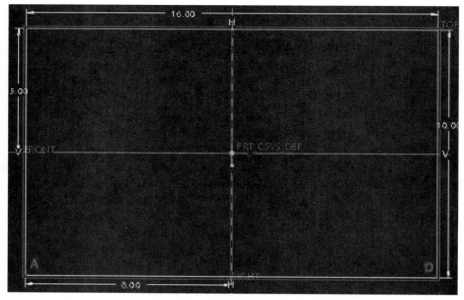

Figure 9.23

Figure 9.24

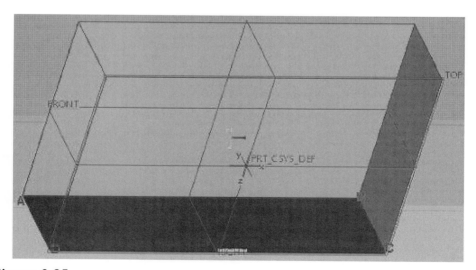

Figure 9.25

2. Click  for visual clarity and sketch the section indicated in Figure 9.26.

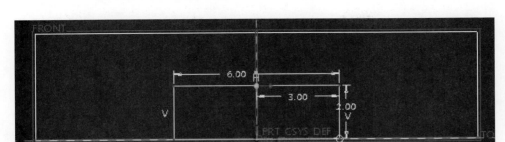

Figure 9.26

3. Click continue ✔ and set the thickness of the section to **4.00** units. Toggle the materials removal on and extrude into the *BOX* part. The dashboard setting should be like that shown in Figure 9.27.

Figure 9.27

4. Turn the Shading mode on ⬜ and preview the part 👁 so that it looks like that shown in Figure 9.28 in the standard orientation.

5. Complete the feature ✔.

6. In the Model Tree window, rename the feature *CUT*.

MODIFYING THE PART

Assume now that it is decided to make surface ABCD curved. We can edit the *BOX* section and make it curved. However, since the box feature was created with the flat surface, as a reference, Pro/ENGINEER will fail in the part generation. Let us try this out.

1. Select the *BOX* in the Model Tree → **Edit Definition** → ⬚.

  (Version 2.0) Select Placement → ▭ Edit... → **Sketch** → **Close**.

2. Select line AD (Figure 9.29a) so that it turns red → Right-click → **Delete** → **Yes**.

  Pro/ENGINEER warned in the previous step because the entity we are about to delete is a reference for another entity. That is another feature is dependent on the curve we are about to delete. Deleting the curve thus makes it impossible for Pro/ENGINEER to place the dependent entity uniquely.

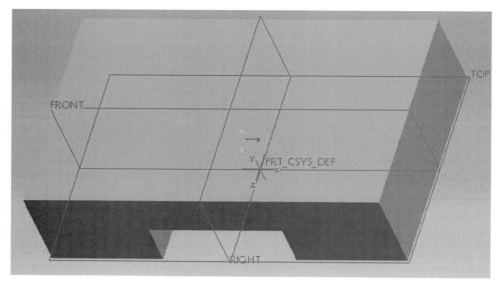

Figure 9.28

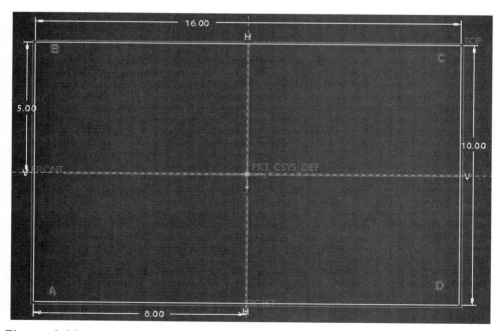

Figure 9.29a

3. Replace the straight line AD with a curved line, so that the section looks like the one drawn in Figure 9.29b.

16.00
H
TOP
5.00
10.00
PRT_CSYS_DEF
V
20.00
8.00

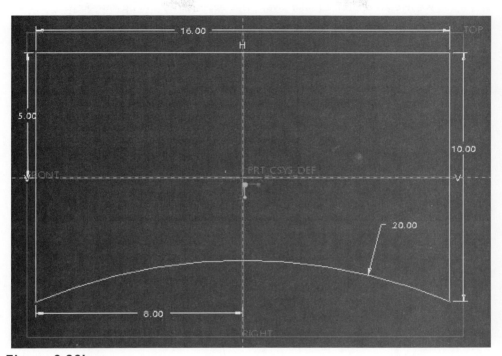

Figure 9.29b

4. Continue the sketch ✔ → **OK** → Complete the feature.

   Pro/ENGINEER fails in regeneration as predicted. The following window (Figure 9.30) is displayed.

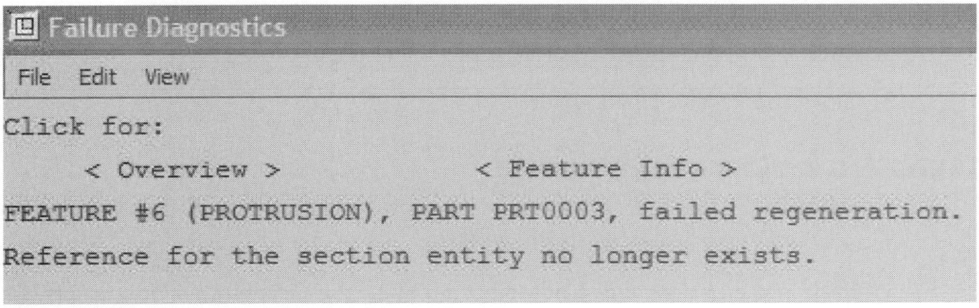

Failure Diagnostics
File Edit View
Click for:
< Overview >
< Feature Info >
FEATURE #6 (PROTRUSION), PART PRT0003, failed regeneration.
Reference for the section entity no longer exists.

Figure 9.30

5. Click **Undo Changes** → **Confirm**.

Part regeneration fails because *CUT* is dependent on curved surface, which we had attempted to take away. This dependency can be viewed using the Parent–Child information.

6. In the Model Tree, right-click on feature box → **Info** → **Parent/Child**. The following window (Figure 9.31) is displayed.

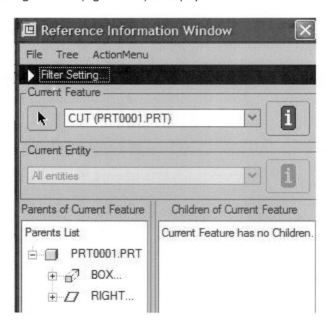

Figure 9.31

7. Observe that feature *CUT* has no children but it has parents. Expand the item BOX in the Parents List, if necessary. Observe that Surfaces are listed as parents. The edge we deleted effectively took away that parent, hence Pro/ENGINEER could no longer place *CUT* and the result was a failure alert!

In order to make the change we are seeking possible, we will need to make feature CUT independent of its present surface parent.

8. Close the Parent–Child window.

CREATING A SKETCHING PLANE

In order to fulfill the task of removing the present parent–child dependency for *CUT*, let us now create a new sketching plane.

1. Click in the background to deselect all. With the Graphics window displayed, click Datum Plane Tool ▱ to create a new plane.

2. Click the FRONT plane in the Model Tree and set the translation to 5.0 to make the new plane coincident with the cut face ABCD (Figure 9.25). The settings of the Datum Plane window are shown in Figure 9.32.

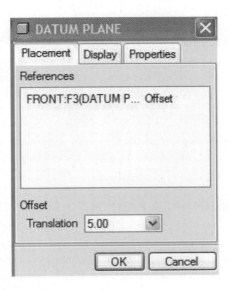

Figure 9.32

3. Click **OK** and observe that the new feature plane DTM1 appears in the Model Tree window.

4. In the Model Tree, select DTM1 and drag before the cut. This makes the new plane appear first before the cut. DTM1 must appear first before CUT in order to make CUT a child of DTM1. The Model Tree now looks like that shown in Figure 9.33.

Edit CUT to make it a child of DTM1, instead of being a child of plane ABCD.

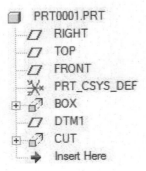

Figure 9.33

1. In the Model Tree, right-click on CUT → **Edit Definition** → ☑.

   (Version 2.0) Select Placement → Edit... → **Sketch**.

2. In the **Placement** window, click DTM1 in the Model Tree as the sketching plane → **Sketch**.

3. Using the standard menu bar, click **Sketch** → **References**.

4. In the References window, right-click on surface reference (SURF5: PROTRUSION) and delete it. The Reference status shows partially placed. We need it to be fully placed by adding another reference to the list of reference entities.

5. Click 🔍 in the same References window.

6. Select the F2(TOP) plane in the Graphics window to include it as a reference for dimensioning. The Reference status shows fully placed.

7. Click Close → ✔ → **OK** → ✔.

INTRODUCING A REQUIRED PART MODIFICATION

Repeat steps 1–4 of the section: *Modifying the part*, above. The modification succeeds! The modification succeeds because the cut surface ABCD is no longer a parent of CUT. Therefore, regardless of changes made to surface ABCD, Pro/ENGINEER can still uniquely place the CUT feature.

The completed part now looks like that shown in Figure 9.34.

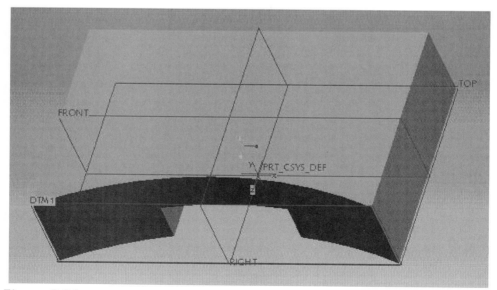

Figure 9.34

EXERCISE

1. In the previous tutorial, modify the part to make surface ABCD flat again.

2. The part shown in Figure Q9.2a and Figure Q9.2b has a hole placed on the curved side. Modify the part to make surface PQRS flat.

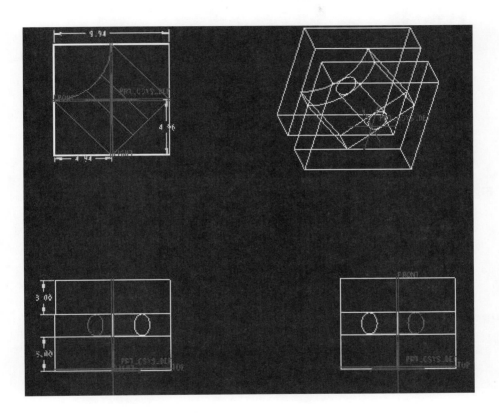

Figure Q9.2a

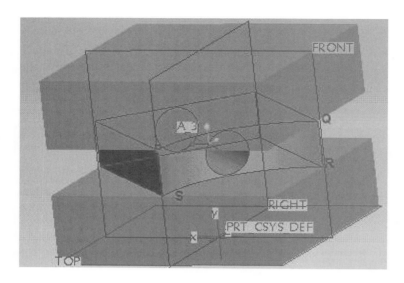

Figure Q9.2b

CHAPTER 10

Relations and Equations

RELATIONS

Relations express dependencies between the dimensions of a feature. Use of relations enable interdependence of dimensions as the following example shows.

CREATING A PART

1. Start Pro/ENGINEER. Create a new part and name the part *relations1*.

2. Using the standard menu bar, click **Insert** → **Extrude**. Go into the Sketcher environment. Drawing on the TOP plane and accepting the defaults, create the profile shown in Figure 10.1.

The numeric representation of dimensions are shown in Figure 10.1. In order to carry out relations, we need to represent these dimensions as symbolic representations.

1. In the standard menu bar, click **Info** → **Switch Dimensions**. Symbolic dimensions like those shown in Figure 10.2 appear.

   The parameters seem to be in the order the elements of the sketch were created and may vary from user to user. We will come back to the use of these parameters in the use of relations shortly.

2. Click ✔ and set the thickness to **2**.

3. Complete the feature ✔ so that the final entity looks like Figure 10.3 in standard orientation.

   Using the Model Tree, right-click on the extruded protrusion.

   Notice that the dimensions are shown as numbers. We want to display the dimensions in terms of symbolic representations.

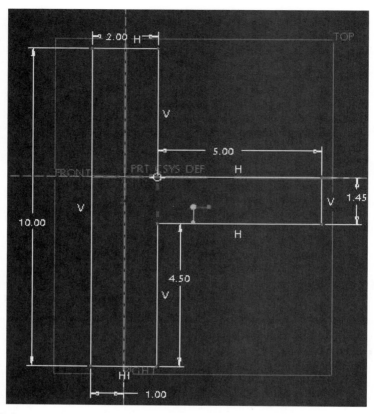

Figure 10.1

4. Using the standard menu bar, select **Info → Switch Dimensions**. The dimensions are displayed as shown in Figure 10.4.

5. Using the standard menu bar, select **Tools → Relations**.

6. Type the relations shown in the **Relations** window that then pops up, as shown in Figure 10.5. Note that the parameters used in the equations depend on what Pro/ENGINEER actually displays, as shown in Figure 10.5. The requirements here are:

   A. Make the arms of the Tee symmetric.

   B. Make the vertical portion of the Tee equal to an arm-length.

   These requirements are met by the equations in the Relations window, Figure 10.5.

   Notice that comments are indicated in the Relations window by preceding a line with /*. Comments are not processed by software but act as a note to the designer or reader as to what the equations do.

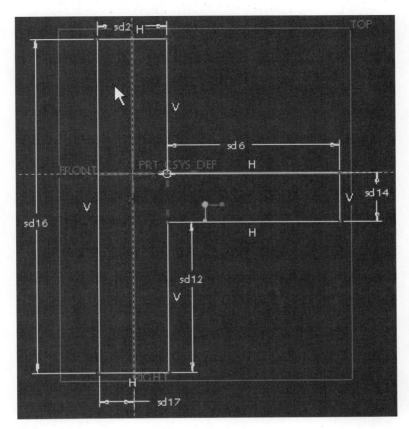

Figure 10.2

As mentioned previously, the actual symbols used vary from display to display. Since the symbols are automatically generated by Pro/Engineer, it is necessary to match appropriate symbols, as displayed on the user's Graphics window. Presently, these symbols represent dimensions as shown in Figure 10.4.

7. Click **Verify** in the Relations window ☑, to ensure that the relations are properly formulated → **OK** → **OK**.

Note: Nothing seems to have happened. To see the effect of the changes we have made, we have to regenerate the model display.

8. Click regenerate model in the standard tool bar ⟳.

The modified part now looks like that shown in Figure 10.6. Notice that the Tee head is now symmetric and that the vertical portion of the Tee is now equal to

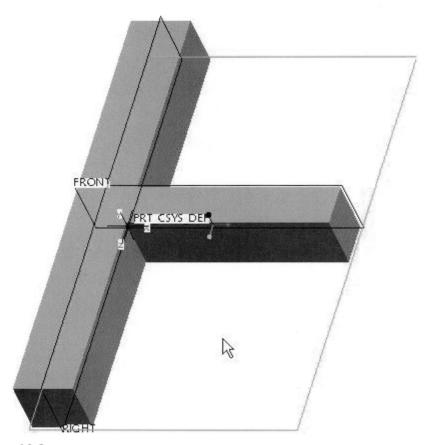

Figure 10.3

an arm-length. It does not end there, the relations we have set up can be reapplied, as the following few steps show.

9. Using the standard menu bar, switch the dimensions to numeric: **Info → Switch Dimensions**.

10. In the Model Tree, right-click on the protrusion → **Edit Definition**.

11. Change the length of the T head, shown in Figure 10.7, from **10** to **15**. Accept changes and return to the model window. Regenerate the model ![icon]. The part now looks like that shown in Figure 10.8.

Notice that the rest of the sides change appropriately, for example the arms are still symmetric. That is because the dimensions are now bound by relations. Changes occur according to the defined relations, and these relations are enforced even when a modification is made.

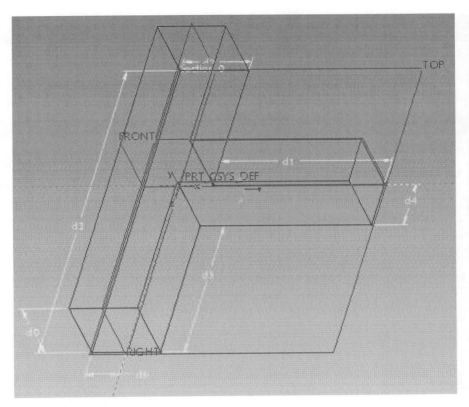

Figure 10.4

EQUATIONS

Equations can be used in defining a profile. Pro/ENGINEER achieves this by using datum curves defined by equations. Equations in Pro/ENGINEER require some knowledge of parametric equations. But do not worry too much about this for now. We will accomplish some expertise through practice.

SKETCHING THE SECTION OF AN EXTRUDED PART USING AN EQUATION

1. Start Pro/ENGINEER and begin a new part → Name the part *Equations1* → Accept the default.

2. **Insert a datum curve** 〰 → **From Equation** → **Done**.

3. Select the **default coordinate system** in the Graphics window .

4. Select cylindrical and type the equation shown in Figure 10.9.

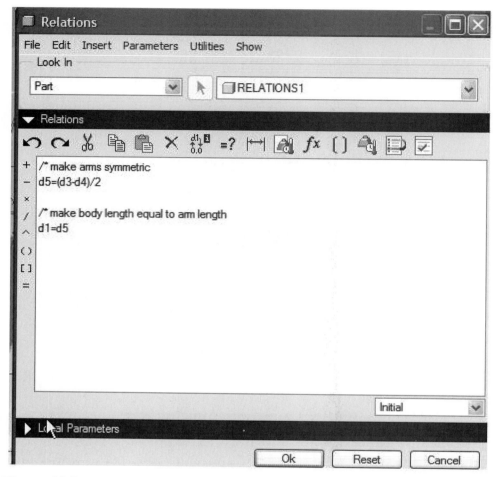

Figure 10.5

5. In the **Notebook** window that opens, click **File → Save**. Close the Notebook window.

The **Curve** window shows that the coordinate system, the type, and the equation are now described as shown in Figure 10.10. If you find the dialog window hidden – and this often happens – activate it by selecting it from the standard menu bar window or by using short cut Ctrl + A.

6. Click **OK**.

The Graphics window now shows the drawn part, as shown in Figure 10.11.

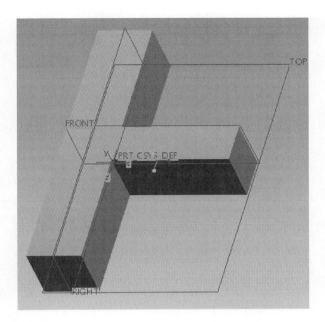

Figure 10.6

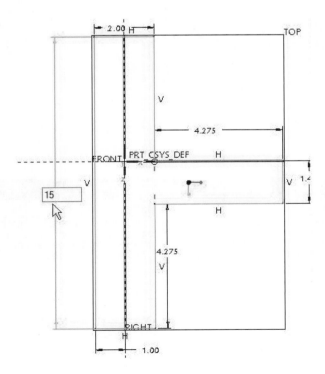

Figure 10.7

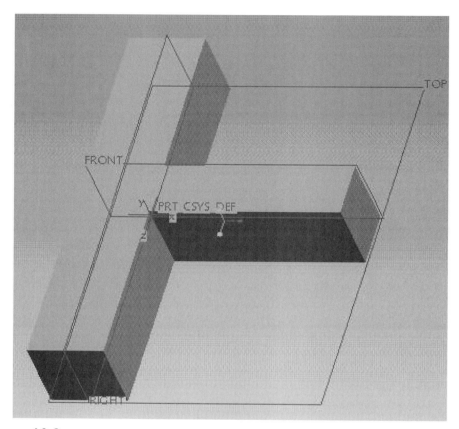

Figure 10.8

```
rel.ptd - Notepad

File  Edit  Format  View  Help
/* For cylindrical coordinate system, enter parametric equation
/* in terms of t (which will vary from 0 to 1) for r, theta and z
/* For example: for a circle in x-y plane, centered at origin
/* and radius = 4, the parametric equations will be:
/*         r = 4
/*       theta = t * 360
/*         z = 0
/*--------------------------------------------------------------

r=3+0.5*((1+cos(180*t))-0.45*(1-cos(360*t)))
theta = 180*t
z = 0
```

Figure 10.9

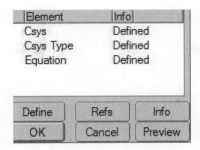

Figure 10.10

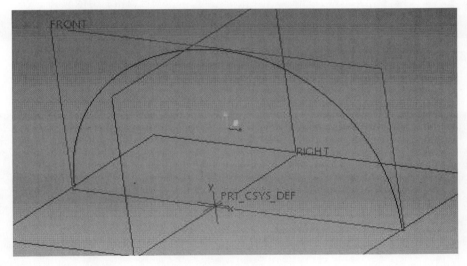

Figure 10.11

We would like to use this profile to be part of our extrude cross section. First of all, notice that the profile is drawn on the FRONT plane.

7. Click Extrude 🔲 → 🖉.

    (Version 2.0) Select Placement → ⟮ Define... ⟯. → Select the FRONT plane as the sketching plane → **Sketch** → **Close**.

The sketcher now appears as shown in Figure 10.12.

8. Draw a line to join the ends of the profile.

9. Although the entity looks closed, the curve is not yet part of our sketch. Therefore, click 🔲 → Click the profile → **Single** → **Close**, so that the sketch now looks like that shown in Figure 10.13.

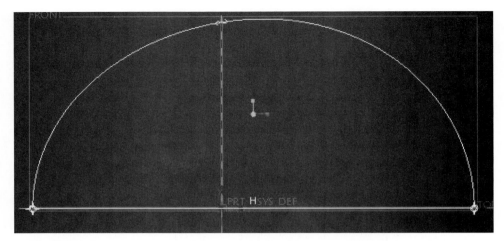

Figure 10.12

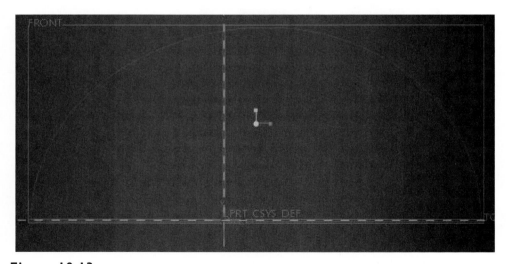

Figure 10.13

10. Continue the sketch ✓ and choose a height of **3** in the dashboard → ✓ → Ctrl + D.

The completed part now looks like that shown in Figure 10.14.

Notice that the cross section is not a semicircle as it appears, but defined by parametric equations.

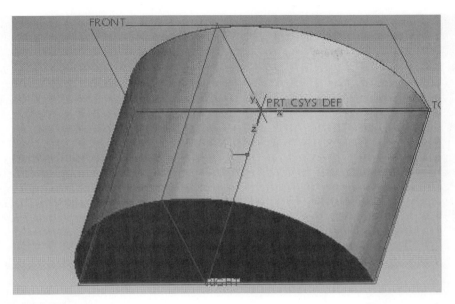

Figure 10.14

SKETCHING THE TRAJECTORY OF A SWEPT SURFACE USING AN EQUATION

1. Start Pro/ENGINEER and begin a new part → Name the part *equations2* → Accept the default.

2. **Insert a datum curve** 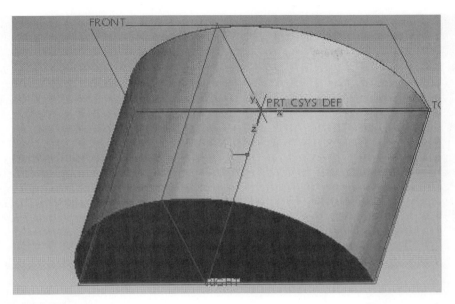 → **From Equation** → **Done**.

3. Select the default coordinate system in the Graphics window .

4. Select **Cylindrical** and type the equation shown in Figure 10.15.

5. In the **Notepad** window, select File → Save → File → Exit.

6. Click **OK** → Ctrl + D, to view in standard orientation.

   The curve generated is as shown in Figure 10.16.

   We are now going to use this curve as a trajectory for a surface sweep. We will create a simple line as the profile to sweep around this trajectory. From the display above, it will appear most appropriate to sketch our profile on the RIGHT plane, which appears perpendicular to the trajectory.

7. Click the **Variable Section Sweep Tool** → Go to the Sketcher.

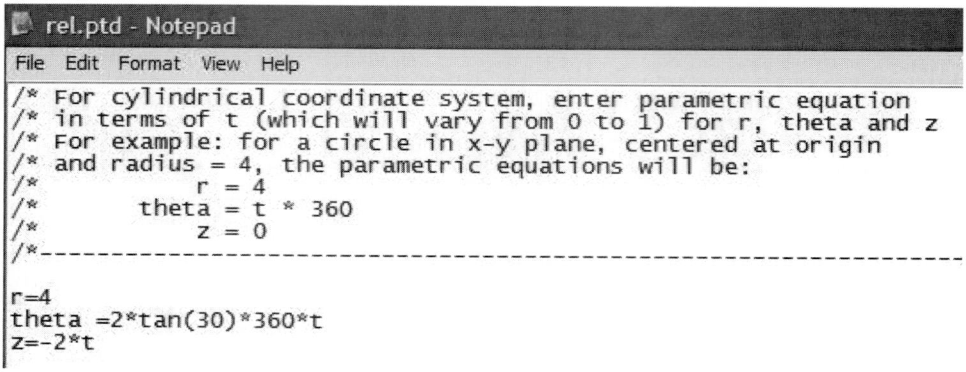

```
rel.ptd - Notepad
File  Edit  Format  View  Help
/* For cylindrical coordinate system, enter parametric equation
/* in terms of t (which will vary from 0 to 1) for r, theta and z
/* For example: for a circle in x-y plane, centered at origin
/* and radius = 4, the parametric equations will be:
/*            r = 4
/*        theta = t * 360
/*            z = 0
/*-----------------------------------------------------------------

r=4
theta =2*tan(30)*360*t
z=-2*t
```

Figure 10.15

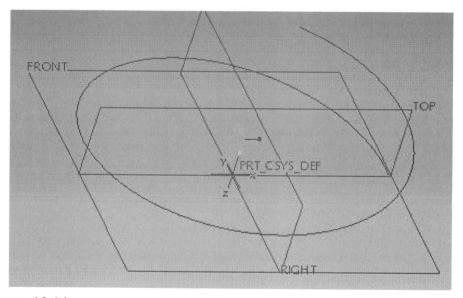

Figure 10.16

However, Pro/ENGINEER will automatically orient the drawing to the start of the datum, as shown by the intersection of the two golden lines.

8. Draw the line AB on the horizontal reference line as shown in Figure 10.17.

9. Click ✔ → Ctrl + D.

The profile looks like that shown in Figure 10.18.

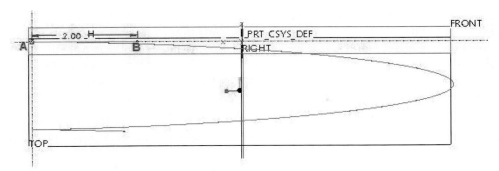

Figure 10.17

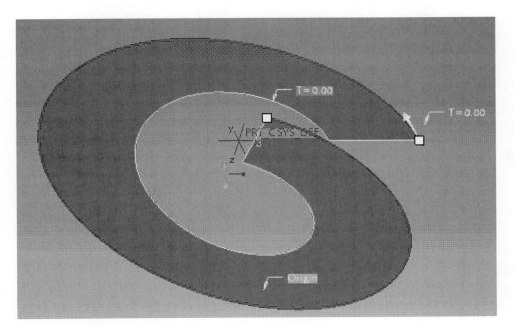

Figure 10.18

10. Complete the feature.

11. Save the file → Close Window.

CREATING A POINT

By now, you can create simple shapes in Pro/ENGINEER. Occasionally, you will need to create a point, for example to define a 3D curve or to locate the position of a load application.

The Datum Point Tool ⸬. This tool can be used to create a point in the sketcher environment or on a feature. It is convenient to use this tool to place a point on, or offset from, a surface. Placing a point on a surface is not sufficient, since the point can be anywhere on that surface. It is necessary to precisely locate the point on the surface by using the **Offset references** settings to locate the point on the surface.

The **Offset Coordinate System Datum Point Tool** ⸬. This tool may be conveniently used when the coordinates in the three dimensions are known. The steps are demonstrated below:

1. Start a new Pro/ENGINEER file and name it *datum_points*.

2. On the tool bar normally to the right-hand side of the Graphics window, select ⸬ ▸ ⸬ ⸬ ⸬ ⸬ . The **Offset Datum Point Tool** window appears. Notice that the dashboard now asks for the coordinate system. This is the coordinate system upon which the created point(s) will be based.

3. Select the default coordinate system /PRT_CSYS_DEF.

4. In the **Offset Datum Point Tool** window, click under name and edit the name and the numeric values of the coordinates.

5. Observe that the created points are displayed on the Graphics window. If the points are not visible, verify that on the standard tool bar the **Datum Point Tool** ⸬ icon is toggled on.

EXERCISE

1. Create a part, shown in Figure Q10.1, whose cross section is defined by a concentric ellipse and circle. The ellipse has minor and major axes of **4** and **5**, respectively, while the radius of the circle is **3** units.

2. A. Open the part created in Figure 10.18. In the Model Tree, click the curve used to create the surface → **Edit Definition** (Figure Q10.2a).

   B. You are back to the specification window. Select the **Equation** → **Define**.

   C. Experiment with different values of r, ϕ, and z to gain some experience with the parametric nature of this curve.

   D. Create the part resembling that shown in Figure Q10.2b.

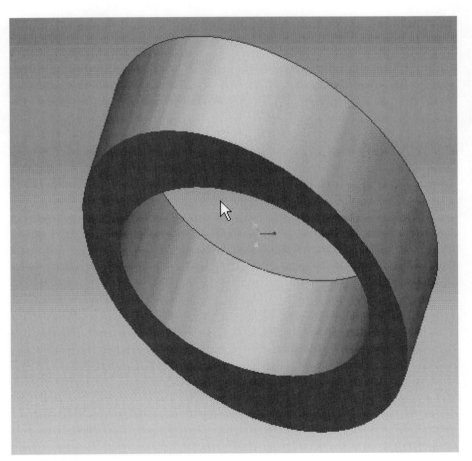

Figure Q10.1

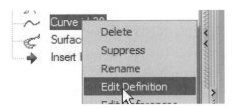

Figure Q10.2a

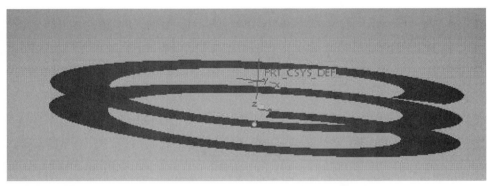

Figure Q10.2b

CHAPTER 11

Parametric Design, Family Tables

PARAMETRIC DESIGN

Parametric design enables the creation of a feature in terms of pre-assigned variables. The variables are assigned values and then the dimensions are expressed in terms of these variables. The advantage of this approach is the ease with which a feature may be modified by simply changing the variable values. Sure, it uses some relations, as discussed in the previous chapter. Beyond relations, parametric design essentially involves a table of parameters that is completed like a form. Based on the form information, a part is automatically generated.

1. Start Pro/ENGINEER and name the part *parameters1*.

2. Select the **Revolve Tool** ⚙ → Sketch ▤.

    (Version 2.0) Select Placement → ⬚ Define... .

3. Select TOP as the sketching plane → **Sketch** → **Close**.

4. Draw a centerline along the vertical plane RIGHT using the line option ⁞.

5. Draw the section shown in Figure 11.1. Ensure that the left vertical line of the section is also drawn.

6. Click ✔ → ✔.

7. View in the standard orientation Ctrl + D.

    The part now looks like that shown in Figure 11.2.

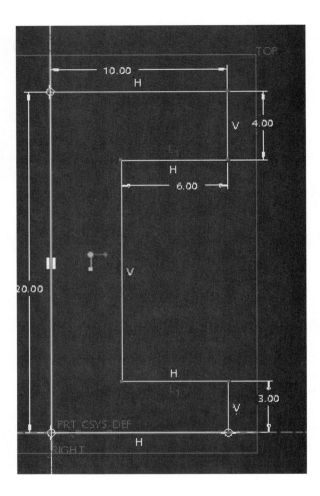

Figure 11.1

8. In the Model Tree, right-click on the revolved feature → **Edit**.

   Notice that the dimensions of the features are displayed. The display may be numeric or symbolic.

9. If the display of the dimensions is not symbolic, then in the standard menu bar, select **Info** → **Switch dimensions**. Note the symbolic representations of the dimensions. They look like the drawing in Figure 11.3, but could be different in yours.

   We can change the dimensions by designing the feature in terms of parameters. This is the essence of parametric modeling.

10. Using the standard menu bar, select **Tools** → **Parameters**. The window shown in Figure 11.4 is displayed.

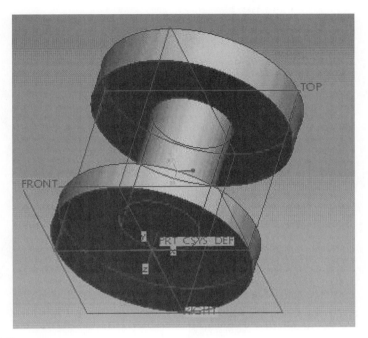

Figure 11.2

11. In the **Parameters** window, click ⊞ to add a new parameter. Add the *length* and *diameter* parameters as shown in Figure 11.5. Note that when the plus sign is clicked, a default name appears in the Parameter window, type *length* to add the length parameter. Repeat the process for the *diameter* parameter. Set the parameter values as shown in Figure 11.5. Click **OK** to accept these parameters.

We now want to bind the dimensions to the parameters we have just created. We use the **Relations** window to accomplish this.

12. Using the standard menu bar, select **Tools** → **Relations**.

We are going to define the symbolic dimensions noted in Step 10 in terms of the parameters defined in Step 11 above.

Suppose we want to generate a part like that shown in Figure 11.3, so that the disk radius (d1 here, may be different in your own display) is one-half of the diameter we entered in Figure 11.5, and the shaft length (d2 here, may be different in your own display) is one-half the length we specified in Figure 11.5, we specify these requirements in the Relations window.

13. In the **Relations** window shown in Figure 11.6, type the following equations.

14. Click **OK** to accept the relations → **Regenerate Model** ⬚. The feature is now transformed to that shown in Figure 11.7.

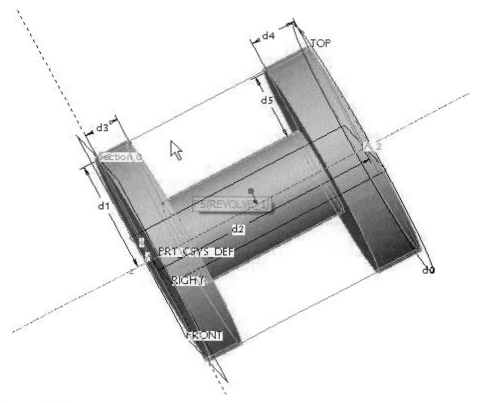

Figure 11.3

15. Using the standard menu bar again, select **Tools → Parameters**. Edit the parameter table as shown in Figure 11.8.

16. Click **OK → Regenerate Model** ⊞. The part is transformed to that shown in Figure 11.9.

    That is it. You have used parameters to control the dimensions of a part. Complex transformations in support of product design can be envisaged.

17. Save the file.

FAMILY TABLES

Family tables enable the automatic generation of parts of same design but of different sizes.

1. You should have the previously created part opened. If not reopen *parameters1*.

2. Using the standard menu bar, click **File → Save a Copy**. Name the copy *familytable1*, as shown in Figure 11.10.

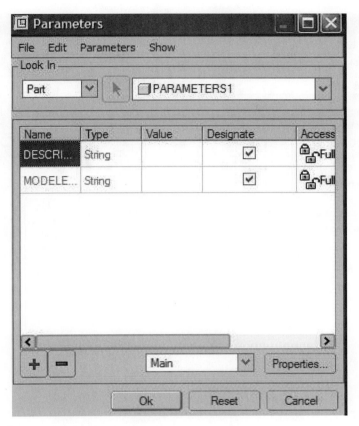

Figure 11.4

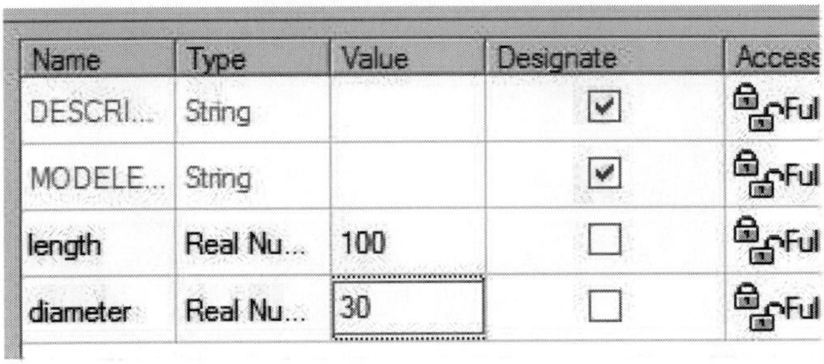

Name	Type	Value	Designate	Access
DESCRI...	String		☑	🔒Ful
MODELE...	String		☑	🔒Ful
length	Real Nu...	100	☐	🔒Ful
diameter	Real Nu...	30	☐	🔒Ful

Figure 11.5

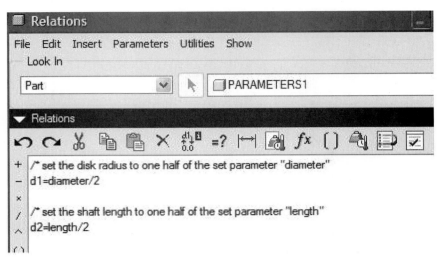

Figure 11.6

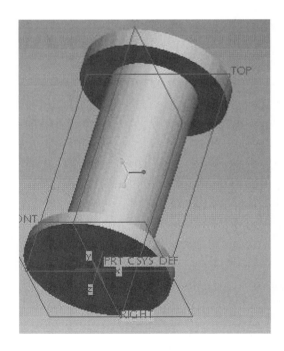

Figure 11.7

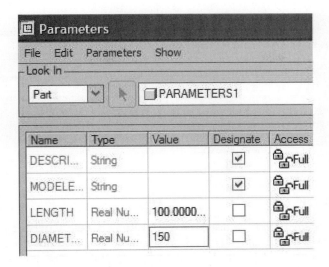

Figure 11.8

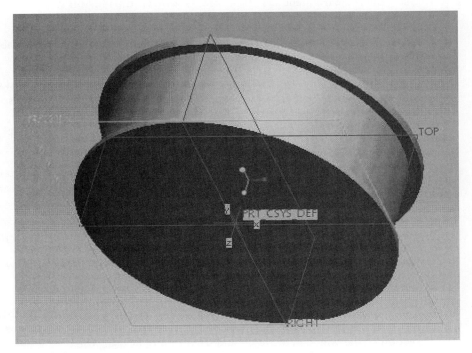

Figure 11.9

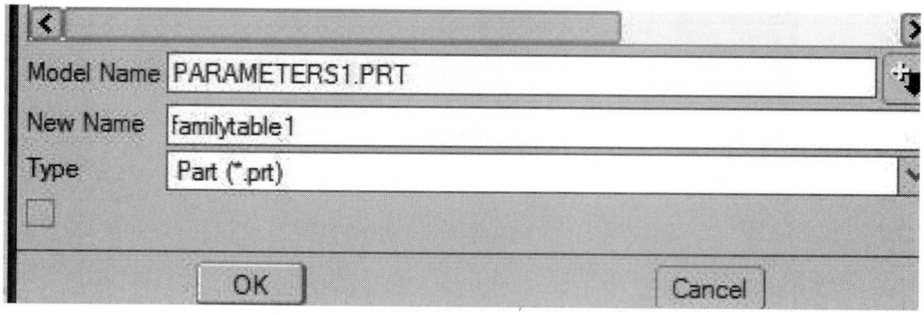

Model Name	PARAMETERS1.PRT
New Name	familytable1
Type	Part (*.prt)

OK   Cancel

Figure 11.10

3. Click **OK**.

4. With *parameters1* file displayed, click **File → Erase Current → Yes**. This removes the *parameters1* file from memory, as we do not need it for this exercise. The file remains saved on the physical storage device, only taken off memory for economic use of the available Random Access Memory (RAM).

5. Using **File → Open**, open the *familytable1* file.

6. In the Model Tree, right-click on the revolved feature → **Edit**.

7. Using the standard menu bar, click **Info → Switch dimension** so that symbolic dimensions are displayed, if necessary.

8. Using the standard menu bar, click **Tools → Family Table**. In the Family Table displayed, click add/delete table columns 🗒.

9. In the Graphics window, select the dimension between the two end disks (d2) as shown in Figure 11.11. Notice that the selected dimension is also displayed in the Family Items window, Figure 11.12. Similarly, select the disk radius d1. These selected dimensions will be used to *drive* our part generation.

 Note: Note that a different symbolic representation may appear in your own display, different from d1 and d2. Just ensure the disk radius and shaft length are the dimensions selected.

10. Observe that the selected dimensions are listed in the Items window. Click **OK**.

11. Create an instance of the part by clicking 🞐 in the **Family Table** window.

12. Create the increments by clicking 🞐 again in the **Family Table** window.

13. The **Patternize instance** window opens up. Click the **Quantity** textbox and enter **4** so that four instances of the part are created. In the **Items** section of the window, move d1 and d2 to the right using the arrow sign. By selecting d1, and d2 in turn, set the variations so that d1 and d2 are each

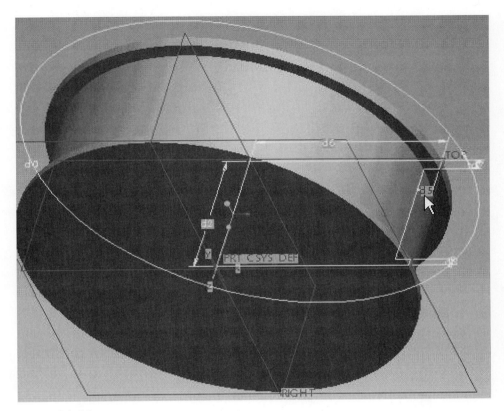

Figure 11.11

incremented by **5** units as shown in Figure 11.13. *Be sure to hit the "Enter" key to accept each increment entry. These increments will be applied on the successive instances that Pro/ENGINEER will automatically generate for you, when required.*

14. Click **OK** → **OK**.

The patterns created exist only in the memory of the computer and are not saved on the hard drive. In order to access it on the hard drive, the part needs to be saved first. Pro/ENGINEER has the information to create the instances of the part.

15. **File** → **Save** → ✔.

16. Go to the working directory and open *familytable1*. The successive patterns created are listed in the **Select Instance** window. These files are automatically named by Pro/ENGINEER as familytable1_ inst0, familytable1_inst1, familytable1_inst 2, and familytable1_inst 3. Notice that the automatic names start with index of zero. Figure 11.14 and Figure 11.15 show the familytable1_ inst0 and familytable1_inst 3 parts.

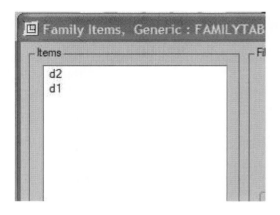

Figure 11.12

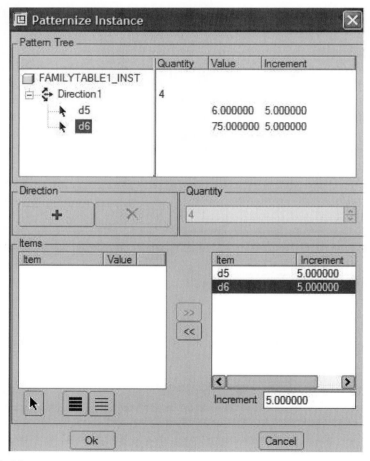

Figure 11.13

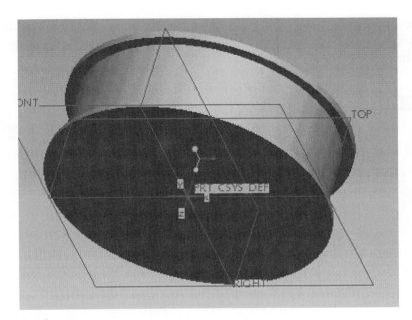

Figure 11.14

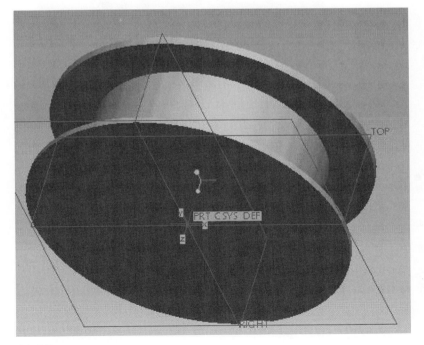

Figure 11.15

Observe how similar parts could be generated automatically. This could be of interest in a manufacturing environment where the customers' needs could vary in one or more of the dimensions of the manufactured component.

EXERCISE

1. Set up a relationship between the diameter and height of the part shown below in Figure Q11.1, so that the height is one-tenth of the part diameter. The lobes are circular and symmetrical. Use free dimensions.

2. Create a family table of the part shown in Figure Q11.1 so that the diameter varies between 10 and 15 and the height remains fixed at 15 units.

Figure Q11.1

CHAPTER 12

Analysis

THE BASICS

Suppose you have drawn a part. By the construction of the part, certain quantities such as weight and volume are associated with the part. Analysis is used to compute such quantities of a feature. Sensitivity analysis may then be used to find a value of any of the dimensions to help realize the desired constraints. An example will make this clearer.

1. Launch Pro/ENGINEER and create a new part named *analysis*, accepting the default. In the features tool bar, select the **Revolve Tool** ⌖. Go to the Sketcher and create, on the TOP plane, a sketch of the dimensions shown in Figure 12.1. Remember to draw a centerline at the left edge, as well as close the sketch by also drawing a line along the left edge. Hence, two lines along the left edge, the centerline, and the closing line, are superposed on the left edge.

2. Accept the sketch → specify a **360** degree revolution → Complete the feature ✔, this generates the feature shown in Figure 12.2.

3. Click 📖 and select the lower surface of the part to be removed during the shell process.

4. Set the shell thickness to **3.75** and complete the feature ✔ to create a feature like that shown in Figure 12.3.

COMPUTING THE VOLUME OF THE SHELL

1. Choose **Insert** → **Model Datum** → **Analysis**. In the **Analysis** dialog box, enter *funnel_vol* as the name of this analysis.

 Note: In the textbox for name (shown in Figure 12.4), type in the name and hit the ENTER key to make sure the system recognizes this entry. This naming enables the

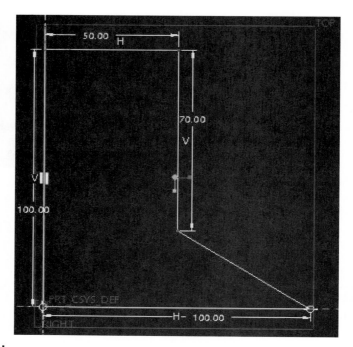

Figure 12.1

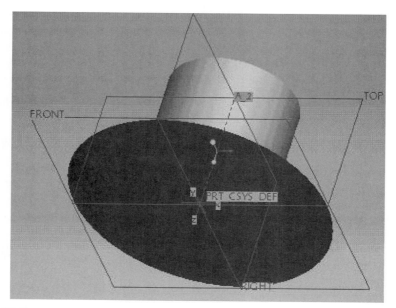

Figure 12.2

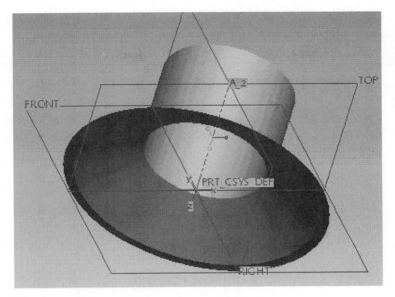

Figure 12.3

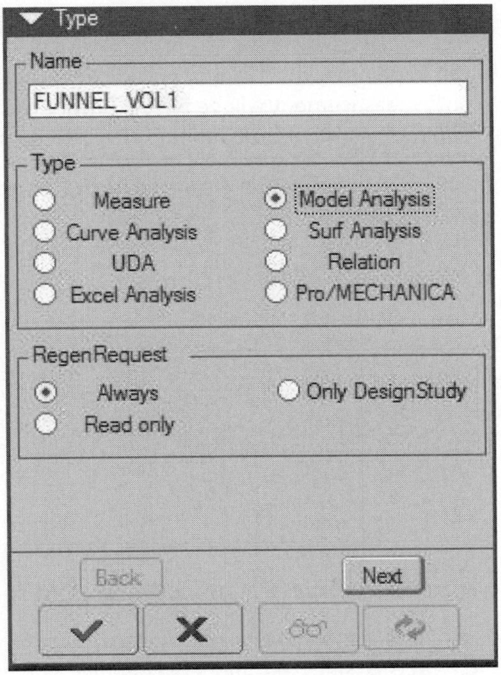

Figure 12.4

analysis to be recalled at a later time. *The Graphics User Interface (GUI) is slightly different in Version 2.0, but the entries are largely similar.*

2. Select Model Analysis as the type of analysis. The rest of the setting should be as shown in the Figure 12.4. Click **Next**.

3. Select One-Sided Volume in the Model Analysis window. This option is seen by clicking the drop-down arrow as shown in Figure 12.5. In the dashboard, Pro/ENGINEER asks for the plane to bound the volume.

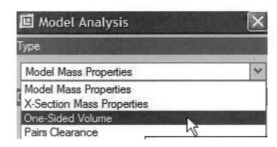

Figure 12.5

4. Select the FRONT plane in the Graphics window to indicate the plane upon which the volume measurement will be based. Ensure that the arrow points into the body by flipping the direction of the arrow, if necessary. Click **OK**.

5. Observe that the model volume is displayed at the bottom of the screen, as shown in Figure 12.6.

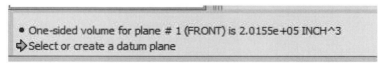

Figure 12.6

6. Press Ctrl + A to activate the window and select **No** to continue the analysis.

7. Click Close. In the **Results Params** window, shown in Figure 12.7, ensure that **Create** is selected and ENTER *one_side_vol1* as the name of the parameter just computed. This param name is to differentiate between

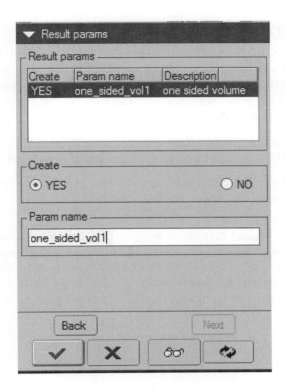

Figure 12.7

different feature parameters that we may compute. Again, it is important to type the name of the parameter in the input box and actually hit the ENTER key for this to work.

8. Click ✔. Observe that the analysis feature is now included in the Model Tree, as shown in Figure 12.8.

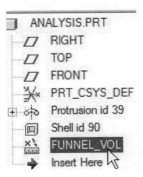

Figure 12.8

VOLUME OF THE CONICAL SECTION

Suppose we want to find the volume of the conical section. We would repeat steps 1–3 above and then define a new plane as passing through the neck of the funnel. We then compute the one-sided volume from the neck to the cylindrical portion of the part. Subtracting the two volumes gives the volume of the conical section. These steps are described as follows.

1. Right-click the revolved feature in the Model Tree → Edit. Observe from the sketch of the feature that the vertical height of the conical section is **30** units.

2. Repeat steps 1–3 of the last section: Computing the volume of the shell. Name the analysis *funnel_vol2*.

3. When Pro/ENGINEER asks for the datum plane (read the dashboard), click **Insert** → **Model Datum** → **Plane**. In the Datum Plane window, ensure that Placement is selected and click the FRONT plane in the Graphics window. Enter −**30** units as offset distance in the **Datum Plane** window (Figure 12.9) → Click **OK**. Ensure that the arrow points into the feature by flipping the arrow, if necessary. The resulting feature is as shown in Figure 12.10.

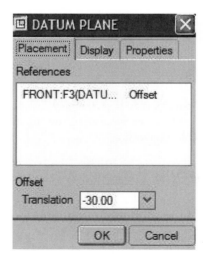

Figure 12.9

4. Notice that the negative offset translation is necessary. The normal to the FRONT plane, shown by the cyan color, is pointing downwards, away from the cylindrical end, whereas we want our offset translation to be going towards the cylindrical portion.

5. Flip the arrow so that it points in the conical section of the funnel, if necessary. Click **OK** and observe that the volume of the shell is displayed in the dashboard as shown in Figure 12.11.

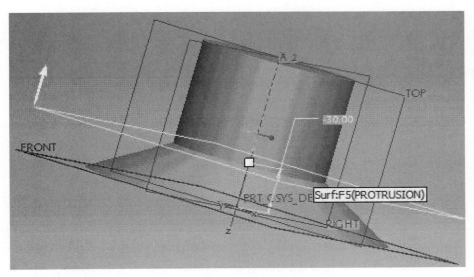

Figure 12.10

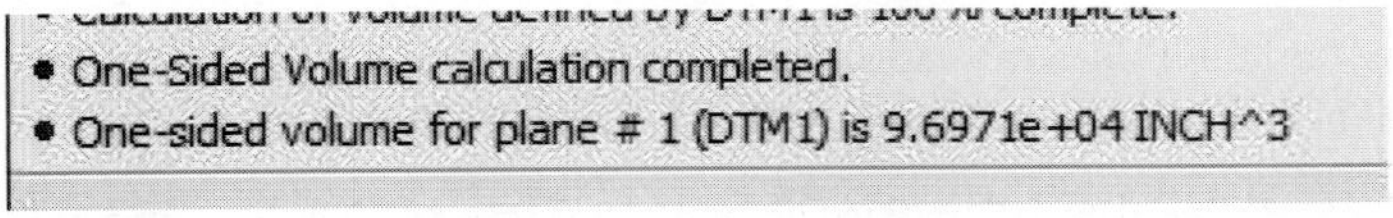

- ~~Calculation of volume defined by DTM1 is 100 % complete.~~
- One-Sided Volume calculation completed.
- One-sided volume for plane # 1 (DTM1) is 9.6971e+04 INCH^3

Figure 12.11

6. The displayed volume in Figure 12.11 is the one-sided volume bounded by the plane DTM1 and the rest of the part in the direction of the pointed arrow. The volume of the cylindrical section may then be obtained by subtracting the two volumes in Figure 12.6 and Figure 12.11.

7. Click ✔ to accept the analysis.

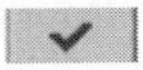

SENSITIVITY ANALYSIS

Suppose we now wish to design the feature so that the volume of the shell is 21,000 cubic units. For this purpose, we will perform a sensitivity analysis.

1. Using the standard menu bar, click **Analysis → Sensitivity Analysis**.

2. In the **Sensitivity** window that opens up, enter *vol1_sensitivity* as the study name, as shown in Figure 12.12.

3. For the variable selection in the Sensitivity window, click ⬚ Variable Selection / Dimension → Click the feature in the Graphics window and select the larger diameter of **100** units.

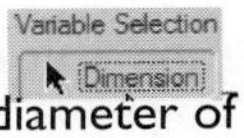

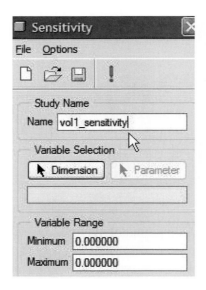

Figure 12.12

Pro/ENGINEER automatically fills in a range of this variable in this case between **90** and **110**, as shown in Figure 12.13.

4. Click **Parameters to Plot** in the **Sensitivity** window, and select *funnel_vol1* → Click **OK**.

5. Keep the settings at **10**, so that the configuration looks like that shown in the sample Sensitivity window in Figure 12.13.

6. Click **Compute**. Pro/ENGINEER generates the graph of the funnel volume with the major diameter of the cone, as shown in Figure 12.14.

The graph of Figure 12.14 shows that by setting the major diameter of the conical section to **104**, all other variables being kept at the present values, we get our desired funnel volume of **210,000** units.

WATCHING THE EFFECTS OF THE CHANGES ON THE MODEL

1. In the **Sensitivity** window, click **Options** → **Preferences**.

2. Check the **Animate model** box → Click **OK**.

3. Arrange the windows so that both the Model Graphics window and the **Sensitivity** window can simultaneously be seen, as shown in Figure 12.15.

4. Click compute and watch the changes applied to the model.

5. In the **Sensitivity** window, change the range of the major diameter to be between **40** and **120**, as shown in Figure 12.16 → Click **Compute**.

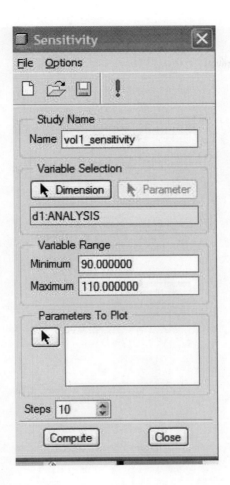

Figure 12.13

Have fun watching Pro/ENGINEER show real-time sensitivity analysis of the model as it goes from an acute to an obtuse orientation and have fun using Pro/Engineer Wildfire!

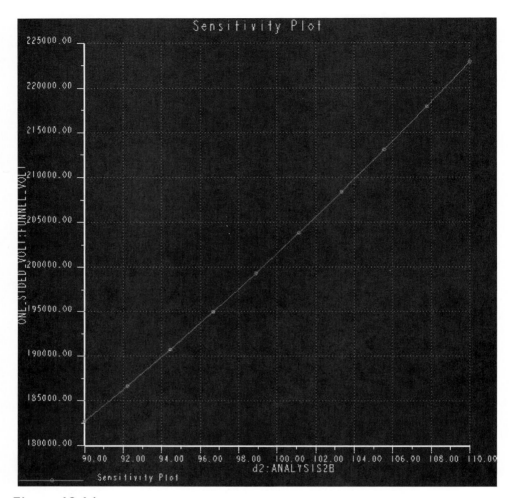

Figure 12.14

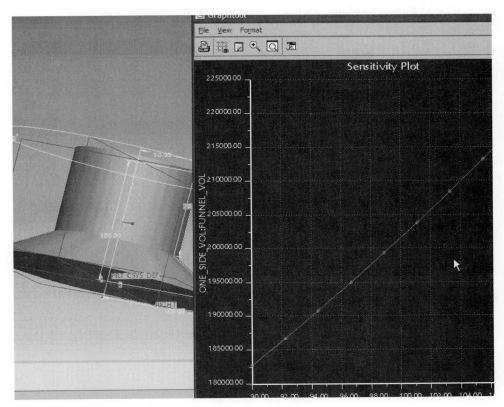

Figure 12.15

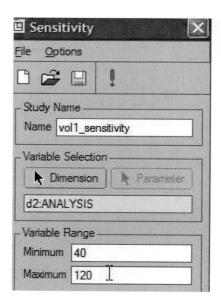

Figure 12.16

EXERCISE

1. A. Create a solid whose sketch is shown in Figure Q12.1a.

   B. Revolve it to form a solid, and then shell it to have a thickness of 1 unit, to form the flask shown in Figure Q12.1b.

2. Find the volume of the flask in Figure Q12.1b.

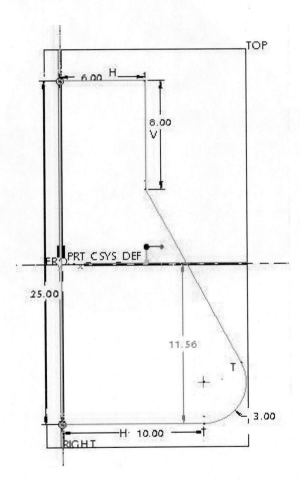

Figure Q12.1a

Figure Q12.1b

INDEX